Stability of Reaction
and
Transport Processes

PRENTICE-HALL INTERNATIONAL SERIES
IN THE PHYSICAL AND CHEMICAL ENGINEERING SCIENCES

NEAL R. AMUNDSON, EDITOR, *University of Minnesota*

ADVISORY EDITORS

ANDREAS ACRIVOS, *Stanford University*
JOHN DAHLER, *University of Minnesota*
THOMAS J. HANRATTY, *University of Illinois*
JOHN M. PRAUSNITZ, *University of California*
L. E. SCRIVEN, *University of Minnesota*

AMUNDSON *Mathematical Methods in Chemical Engineering:
Matrices and Their Applications*
AMUNDSON AND ARIS *Mathematical Methods in Chemical Engineering, Volume 2,
First Order Partial Differential Equations with Applications*
ARIS *Elementary Chemical Reactor Analysis*
ARIS *Introduction to the Analysis of Chemical Reactors*
ARIS *Vectors, Tensors, and the Basic Equations of Fluid Mechanics*
BALZHISER, SAMUELS, AND ELIASSEN *Chemical Engineering Thermodynamics*
BRIAN *Staged Cascades in Chemical Processing*
CROWE ET AL. *Chemical Plant Simulation*
DENN *Stability of Reaction and Transport Processes*
DOUGLAS *Process Dynamics and Control, Volume 1, Analysis of Dynamic Systems*
DOUGLAS *Process Dynamics and Control, Volume 2, Control Systems Synthesis*
FOGLER *The Elements of Chemical Kinetics and Reactor Calculations:
A Self-Paced Approach*
FREDRICKSON *Principles and Applications of Rheology*
FRIEDLY *Dynamic Behavior of Processes*
HAPPEL AND BRENNER *Low Reynolds Number Hydrodynamics:
with Special Applications to Particulate Media*
HIMMELBLAU *Basic Principles and Applications and Calculations in
Chemical Engineering, 3rd Edition*
HOLLAND *Fundamentals and Modeling of Separation Processes:
Absorption, Distillation, Evaporation, and Extraction*

HOLLAND *Multicomponent Distillation*

HOLLAND *Unsteady State Processes with Applications in Multicomponent Distillation*

KOPPEL *Introduction to Control Theory with Applications to Process Control*

LEVICH *Physicochemical Hydrodynamics*

MEISSNER *Processes and Systems in Industrial Chemistry*

MODELL AND REID *Thermodynamics and Its Applications in Chemical Engineering*

NEWMAN *Electrochemical Systems*

OHARA AND REID *Modeling Crystal Growth Rates from Solution*

PERLMUTTER *Stability of Chemical Reactors*

PETERSON *Chemical Reactor Analysis*

PRAUSNITZ *Molecular Thermodynamics of Fluid-Phase Equilibria*

PRAUSNITZ, ECKERT, ORYE, AND O'CONNELL *Computer Calculations for Multicomponent Vapor-Liquid Equilibria*

RUDD ET AL. *Process Synthesis*

SCHULTZ *Polymer Materials Science*

SEINFELD AND LAPIDUS *Mathematical Methods in Chemical Engineering, Volume 3, Process Modeling, Estimation, and Identification*

WHITAKER *Introduction to Fluid Mechanics*

WILDE *Optimum Seeking Methods*

WILLIAMS *Polymer Science and Engineering*

PRENTICE-HALL, INC.

PRENTICE-HALL INTERNATIONAL, INC., UNITED KINGDOM AND EIRE

PRENTICE-HALL OF CANADA, LTD., CANADA

Stability of Reaction

and

Transport Processes

MORTON M. DENN

Professor of Chemical Engineering
University of Delaware

PRENTICE-HALL, INC.

Englewood Cliffs, New Jersey

Library of Congress in Publication Data

DENN, MORTON M.
  Stability of reaction and transport processes.

  (Prentice-Hall international series in the physical
and chemical engineering sciences)
  Includes bibliographies.
  1. Chemical reaction, Conditions and laws of.
2. Transport theory. 3. Stability. I. Title.
TP155.D36   660.2′9′9   74–13037
ISBN 0–13–840264–7

10 9 8 7 6 5 4 3 2 1

Printed in the United States of America

PRENTICE-HALL INTERNATIONAL, INC., *London*
PRENTICE-HALL OF AUSTRALIA, PTY. LTD., *Sydney*
PRENTICE-HALL OF CANADA, LTD., *Toronto*
PRENTICE-HALL OF INDIA PRIVATE LIMITED, *New Delhi*
PRENTICE-HALL OF JAPAN, INC., *Tokyo*

Contents

Preface

Stability plays a central conceptual role in many of the areas traditionally studied in graduate programs in engineering, but it is rarely presented to the student as a unified subject with its own foundations and methods. Thus, students specializing in reactor analysis and fluid mechanics, for example, are often unaware of the close connections, and advances in one field are slow in reaching the other. This communication barrier has been particularly true in nonlinear stability.

The book is based on courses which I have taught at the University of Delaware. The complete text forms the basis for a one-semester (forty-hour) course. In some years I have used portions of the material in an advanced fluid mechanics course. The graduate students have all taken a standard engineering mathematics course. They understand elementary matrix manipulations, including the concept of an eigenvalue, and separation-of-variables solution of linear homogeneous partial differential equations. Except for these concepts the text is self-contained. The student who has read the entire book should be prepared to solve problems and read the literature in linear and nonlinear stability for lumped and distributed parameter systems.

I have incurred many debts in the preparation of the book. Portions of the text reflect work done in collaboration with my graduate students J. R. Black, R. F. Ginn, K. C. Porteous, J. J. Roisman, R. Rothenberger, Z.-S. Sun, and G. R. Zeichner, and my colleagues B. E. Anshus, D. H. McCoy, and A. B. Metzner. R. J. Fisher did the calculations in Chapter 5 and read a complete draft for clarity and errors. I have received considerable guidance in reading the relevant literature from my colleagues C. A. Petty and E. Ruckenstein, as well as from R. Aris, D. D. Joseph, D. Luss, R. L. Sani, and R. A. Schmitz, who have provided me with bibliographies, extensive collections of reprints, and prepublication manuscripts. Much of the preparation was done during a sabbatical leave from the University of Delaware during which I held a fellowship from the John Simon Guggenheim Memorial Foundation. I am particularly grateful to the trustees of the University and the Foundation for this support.

MORTON M. DENN

Experimental Foundations

1

1.1 Introduction

The behavior of a large number of physical and chemical processes of engineering interest is dominated by considerations of stability. From an engineering point of view the stability problem can be posed as follows:

> *A system is designed to operate in a given state. During the course of operation it is inevitable that small disturbances will enter. Will the system operate close to its design state despite the disturbances, or will the disturbances cause the system to move away from its design state?*

We shall sharpen definitions later, but it is clear that the first case, in which the design state is maintained, is *stable*. The latter case is *unstable*. The purpose of this book is to define quantitative methods of determining when a system will be stable or unstable, as well as to consider two closely related questions:

1. How *large* a disturbance is required to cause instability?
2. If a state is unstable, what is the system behavior following a disturbance?

The type of behavior which interests us can be visualized schematically in Fig. 1.1, where we consider a ball which can move along a sequence of ridges

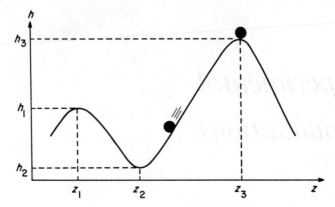

Figure 1.1 Schematic diagram of stable and unstable equilibria.

under the action of gravity, retarded by friction. The elevations h_1, h_2, or h_3, corresponding to locations z_1, z_2, and z_3, respectively, might be design states. If we carefully place the ball down at any of these locations, it will not move. These are *equilibrium*, or *steady-state*, positions for the dynamical system, where no motion takes place and time derivatives are zero.

Clearly, z_1 and z_3 are unstable equilibria, for following any small motion away from the top of the hill, the ball will keep on rolling and not return. z_2 is a stable equilibrium, for following some small displacement the ball will roll back to the bottom of the valley. Even this valley is unstable to a sufficiently large disturbance, however, for if we initially displace the ball a distance greater than $|z_2 - z_1|$ to the left, it cannot return to z_2 but will continue moving to the left.

The rolling ball is a useful model, but, of course, it is of no practical interest. The remainder of this chapter consists of a series of descriptions of experiments which illustrate problems of stability that are of real engineering importance. The examples are drawn from reaction engineering, fluid mechanics, heat transfer, and mass transfer in order to emphasize the breadth of application. Following these descriptions, the remainder of the book deals with the quantitative aspects of predicting the occurrence of such phenomena and analyzing their behavior.

1.2 Chemical Reaction

A chemical reactor is a vessel in which reactants are brought into intimate contact at a fixed temperature for a specified mean time in order to create desired products. Figure 1.2 is a schematic of a continuous flow well-stirred liquid-phase reactor. The stirring ensures a homogeneous concentration and

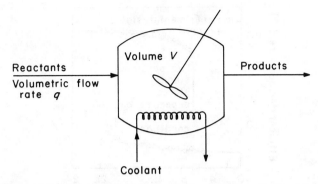

Figure 1.2 Schematic diagram of a continuous flow well-stirred reactor.

temperature distribution within the reactor. The mean time a fluid element spends in the reactor, often called the *residence time*, is

$$\text{residence time:} \quad \theta = \frac{V}{q}$$

The residence time is a primary design variable.

Flow reactors are designed to operate at a steady state, though small fluctuations in flow, coolant properties, feed composition, etc., prevent a total absence of transients. Reactors are sometimes known to "run away." That is, the temperature falls rapidly, stopping the reaction, or, more seriously, the temperature rises rapidly, possibly damaging the equipment or causing safety problems. A runaway reactor is a stability problem, where the system does not return to its steady state following a disturbance but rather moves further away.

There are no good published data on unstable industrial reactors. Several laboratory studies have been carried out, however, which demonstrate some of the important phenomena under controlled conditions. For example, S. A. Vejtasa and R. A. Schmitz studied the reaction between thiosulfate and peroxide

$$2\,\text{Na}_2\text{S}_2\text{O}_3 + 4\,\text{H}_2\text{O}_2 \quad \longrightarrow \quad \text{Na}_2\text{S}_3\text{O}_6 + \text{Na}_2\text{S}_2\text{O}_4 + 4\,\text{H}_2\text{O}$$

in an adiabatic (no cooling) stirred flow reactor. They varied the residence time, θ, for fixed feed conditions and measured the steady-state temperature. Data are shown in Fig. 1.3. At small and large residence times there is a well-defined, unique steady-state operating temperature. In the range roughly $7 \leq \theta \leq 18$, however, *two* distinct steady operating temperatures were obtained, depending on start-up conditions.

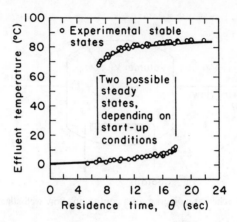

Figure 1.3 Experimental reactor temperature as a function of residence time for the reaction between thiosulfate and peroxide. [After S. A. Vejtasa and R. A. Schmitz, *AIChE J.*, *16*, 410 (1970), with permission.]

The implications of this observation are clear. Suppose that a reactor is designed to operate with $\theta = 12$ sec and a corresponding temperature of about 80°C. This represents a high conversion of thiosulfate. There must be some disturbance such that, following the transient, the reactor will go to a steady-state temperature not of 80°C but rather 4°C, with a correspondingly small conversion. Or, suppose that a reactor is designed for $\theta = 7$ sec, $T = 67$°C. A slight increase in q would reduce the residence time only slightly, but the temperature following the transient would fall to 3°C. These are manifestations of the runaway phenomenon.

A related, quite interesting phenomenon has been studied in laboratory reactors by several workers. Under some conditions a reactor with *constant* feed conditions never achieves a steady state but continues to cycle. Figure 1.4 shows one cycle of data of G. P. Baccaro, N. Y. Gaitonde, and J. M. Douglas for the reaction of acetyl chloride to acetic acid

$$CH_3COCl + H_2O \longrightarrow CH_3COOH + HCl$$

in a cooled continuous flow stirred reactor. Such cyclic reactions are known to occur in some biological systems and may be the mechanism for the "biological clocks" which govern the response of many organisms. Reactor cycling has often been observed in industrial scale systems. The oscillations in such a situation will usually be asymmetric about the design value, leading to an average output over a period which is different from the nominal steady-state design value. Under some conditions this difference might be favorable, leading to a higher average conversion than the steady-state value. It has even been suggested that there may be times when it would pay to install a

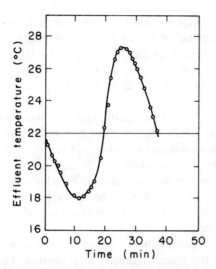

Figure 1.4 Temperature oscillations in the acetyl chloride-acetic acid reaction. [After G. P. Baccaro, N. Y. Gaitonde, and J. M. Douglas, *AIChE J.*, *16*, 249 (1970), with permission.]

control system to *enhance* such fluctuations rather than to try to control the process to produce a time-invariant output.

These phenomena of steady-state multiplicity and runaway and failure to attain any steady state occur in other reaction configurations and are fundamental to the theory of explosions. We shall deal in several chapters in the text with the well-stirred reactor and with the heterogeneous catalyst, where reaction occurs on the surface of a porous solid catalyst with accompanying diffusive transport of mass and energy. These two cases represent important prototypes from which an understanding of other reaction systems easily follows.

1.3 Fluid Flow

In the motion of fluids, transitions between flow regimes occur frequently and need to be accounted for in engineering correlations. The most common such transition occurs in pressure-driven flow in a pipe or between parallel plates. Flow regimes are defined in terms of the dimensionless *Reynolds number,*

$$\mathrm{Re} = \frac{HU\rho}{\mu}$$

H is the pipe diameter or plate spacing, U the mean velocity, ρ the density, and μ the viscosity. For Re < 2100 in pipes and Re < 1300 between plates the streamlines for a Newtonian fluid are parallel and the velocity profile is parabolic. This is *laminar* or *Poiseuille flow*, a configuration which is steady in time and hence represents an equilibrium state for the equations of motion. At higher Reynolds numbers the flow becomes turbulent, indicating that the parabolic profile is unstable and cannot be maintained in the presence of disturbances.

The transition to turbulence is difficult to rationalize without careful examination of the governing equations of motion because of the curious role of the viscosity. One anticipates a tendency of viscous resistance to damp out disturbance motions and to stabilize the laminar flow. Countering this tendency, however, is a subtle mechanism by which the Reynolds stresses transfer energy from the main flow to the secondary motion and support the growth of the instability. We shall consider these effects quantitatively in some detail subsequently.

There are a number of other important flow instabilities which differ qualitatively from the laminar-turbulent transition. Laminar flow of a molten polymer becomes unstable during extrusion at a Reynolds number many orders of magnitude smaller than the critical 2100 for Newtonian fluids, and the subsequent flow appears to be more structured than normal turbulence. This *melt fracture*, or *elastic turbulence*, transition depends on the generation of elastic stress in the molten polymer and correlates with a critical value of the *elastic shear strain, S_R,*

$$S_R = \frac{\tau_w}{G}$$

Here, τ_w denotes the wall shear stress and G the elastic shear modulus of the polymeric liquid. An instability called *draw resonance* limits the drawing of threads or sheets from molten polymers. Here, the elastic stresses are a stabilizing factor. The onset of these instabilities is a limiting factor in many polymer processing operations. We shall show in Chapter 12 how these processing instabilities can be theoretically predicted.

The most structured of the flow transitions, hence the most studied, is that between long rotating concentric cylinders. At low rotational speeds the streamlines are circular and lie in a plane normal to the cylinder axis. At a critical rotational speed the steady circular streamline flow becomes unstable. This instability is manifested by an abrupt change in slope in the torque-rotational speed curve, as shown in Fig. 1.5 for a case in which the inner cylinder is rotating and the outer stationary. The shear rate there is defined as

$$\text{shear rate:} \quad \Gamma \equiv \frac{R\Omega_1}{\delta}$$

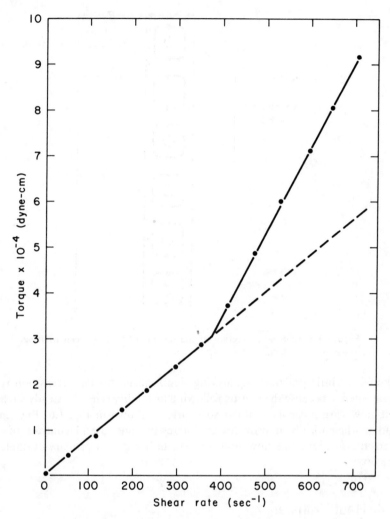

Figure 1.5 Instability in flow of a 60% glycerine in water solution between rotating concentric cylinders, $R/\delta = 28$. [J. J. Roisman, M.Ch.E. Thesis, University of Delaware, Newark, 1968.]

Ω_1 and Ω_2 are the angular velocities of the inner and outer cylinders, respectively; R the radius of the inner cylinder; and δ the spacing between cylinders. It is assumed throughout that $\delta \ll R$. In that case the transition occurs at a critical value of the *Taylor number*

$$T = 2\Omega_1 [\Omega_1 - \Omega_2]\delta^3 \frac{R\rho^2}{\mu^2}$$

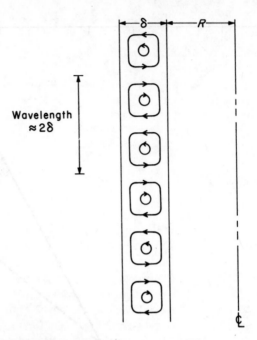

Figure 1.6 Schematic diagram of the secondary flow between rotating concentric cylinders.

Though of little practical engineering significance, this flow transition is of great interest because the motion following instability is itself a steady stream-line flow with a regular cellular structure, as shown in Fig. 1.6. Physically and mathematically it provides an interesting and useful contrast to the transition in Poiseuille flow, and we shall utilize it as a primary example in the text.

1.4 Heat Transfer

Stability plays an important role in heat transfer. The simplest physical situation of interest is one in which a liquid is contained between two large flat plates separated by a distance δ, as shown in Fig. 1.7. The bottom plate is heated. The heat flux Q is just sufficient to maintain the bottom plate at a temperature ΔT above that of the top plate. If conduction through the liquid is the only mode of heat transfer, the dimensionless heat flux, or *Nusselt number*,

$$Nu = \frac{Q\delta}{k_T \Delta T}$$

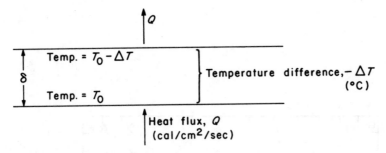

Figure 1.7 Schematic diagram of a heat flux experiment to determine the onset of convective instability.

will have a value of unity. k_T is the liquid thermal conductivity. It is observed, in fact, that for sufficiently large δ or ΔT the Nusselt number deviates systematically from unity, indicating the onset of a new mode of heat transfer.

The qualitative explanation here is not a difficult one. The liquid near the heated bottom plate is less dense than that above it. Thus, if a column of liquid is disturbed, there will be a tendency for the heavier liquid to fall to the bottom, much like a ball at equilibrium at the top of a hill. Such a convective motion would increase the Nusselt number. On the other hand, viscous forces in the liquid will resist the motion. Thus, the convective motion will not begin until the driving force is sufficient to overcome the viscous resistance. For small density differences (small ΔT) and/or small plate spacing the liquid will be quiescent, and only conduction will occur. When these values become large enough convective heat transfer will supplement the conduction.

From considerations of dimensional analysis it can be established that the Nusselt number will depend, in general, on the Rayleigh and Prandtl numbers,

$$\mathrm{Ra} = g\alpha\,\Delta T \delta^3\,\frac{\rho}{\mu\kappa}$$

$$\mathrm{Pr} = \frac{\mu}{\rho\kappa}$$

g is the gravitational acceleration, α the coefficient of thermal expansion, and κ the thermal diffusivity. Furthermore, the dependence of Nu on Pr can be shown to be a second-order one in a nearly quiescent fluid. Figure 1.8 shows data of P. L. Silveston on heat transfer in five different liquids of widely varying physical properties. The data lie on a single line and show a clear transition from conduction (Nu = 1) to convection (Nu > 1) at a value of Ra of approximately 1700. We shall show in Chapter 13 that this observation is in excellent agreement with the theoretical prediction, and in Chapter 16 we shall develop a method which can be used to predict the heat transfer rate beyond the transition.

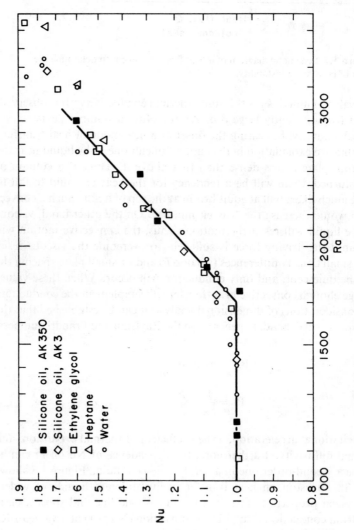

Figure 1.8 Nusselt number as a function of Rayleigh number. Data of P. L. Silveston. [After S. Chandrasekhar, *Hydrodynamic and Hydromagnetic Stability*, Oxford University Press, Inc., New York, 1961, with permission.]

1.5 Mass Transfer

There is an interesting phenomenon in interfacial mass transfer, illustrated in Fig. 1.9, which depends on the stability of an equilibrium state. The data shown here were obtained by D. R. Olander and L. B. Reddy in a continuous flow vessel with a distinct measurable interface between organic and aqueous phases. The rate of mass transfer between phases can be written as

$$r = K_m a(c - c_e)$$

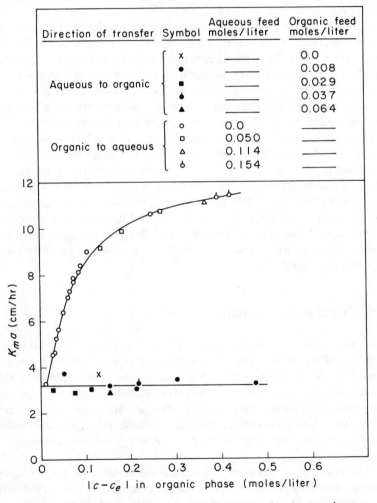

Figure 1.9 Transfer of nitric acid between isobutanol and water phases. [After D. R. Olander and L. B. Reddy, *Chem. Eng. Sci.*, *19*, 67 (1964), with permission.]

where c and c_e are the concentration and equilibrium concentration, respectively, of solute in the organic phase; a is the interfacial area; and K_m the mass transfer coefficient. The data were all taken at identical agitation conditions and phase volumes, so the bulk hydrodynamics were the same for all experiments. When the solute was fed in the aqueous phase and transferred to the organic phase, a constant value of K_m was obtained for all $c - c_e$, as would be expected. When the solute was fed in the organic phase and transferred to the aqueous phase, however, the measured mass transfer coefficient showed a strong functional dependence on $c - c_e$ and increased more than threefold over the previously measured value. Data from the two sets of experiments agree only in the limit of small $|c - c_e|$.

This behavior indicates that the presence of the excess solute in the organic phase induces a flow in the neighborhood of the interface which is absent when the mass transfer is from the aqueous phase, causing the higher mass transfer coefficient. A qualitative explanation of the phenomenon depends on the observation that surface tension is a function of solute concentration. Small local fluctuations in surface concentration will therefore cause variations in surface tension, destroying the stress equilibrium in the interface. The unbalanced stresses will induce a hydrodynamic flow, which is resisted by the viscous forces. As in the case of convective flow induced by a density gradient described previously, there will be a critical condition for which the surface forces overcome the viscous forces and flow begins. This point will depend on the concentration differences in the system, the concentration dependence of surface tension, and the viscosities and solute diffusivities of each phase. We shall deal with the quantitative aspects of this phenomenon, known as the *Marangoni effect*, in Chapter 14, though not in as complex a situation as the experiments described here.

1.6 Concluding Remarks

The reaction and transport processes described in the preceding sections were chosen to illustrate the effect of instability of an equilibrium in a variety of situations. We shall deal with each of them in one or more places in the text to illustrate how some general principles of stability analysis are applied in a wide variety of engineering situations. The primary purpose of the text is to show that a small number of basic quantitative principles are used in all stability analyses, regardless of the physical or physicochemical source of the instability.

The first half of the book focuses on the stability of *lumped parameter systems*, whose dynamical behavior can be described by ordinary differential equations. For illustrative purposes we shall most often use the stirred chemical reactor described in Sec. 1.2, and we shall determine methods for examining the multiplicity of steady states, instability of the steady state, and sustained oscillations. We shall also consider for illustration in these chapters the stability of feedback control systems and transition to non-Newtonian behavior in a rheological model of polymer solutions.

The second half of the book focuses on the stability of *distributed parameter systems*, whose dynamics are described by partial differential equations. Here, we shall consider for illustration the stability of a heterogeneously catalyzed chemical reaction and the fluid mechanics, heat transfer, and mass transfer processes described in the preceding sections of this introduction. The mathematical tools utilized in the latter half are based on procedures first developed for the simpler lumped parameter systems.

It is appropriate at this point to say a word about scope and limitations. We have set out to cover the complete field of linear and nonlinear approaches to stability, using examples from a wide variety of applications areas. The compromise to allow us to do this has been depth of coverage in any particular area. Thus, we shall consider only the simplest reaction kinetics for the flow reactor and only the simplest catalyst geometry in heterogeneous catalysis. We shall not deal with the establishment of structure in multicomponent diffusion and reaction, a problem with far-reaching implications in cell biology. We shall consider only certain fluid mechanical applications and shall neglect such interesting and important phenomena as the formation of ripples on a falling film and the breakup of a liquid jet into droplets. The principles and techniques all carry over to these and other problems, and we shall provide a guide to the relevant literature in the Bibliographical Notes at the end of each chapter and in the Postface.

BIBLIOGRAPHICAL NOTES

The phenomenon of multiple steady states in a flow reactor seems to have been first suggested by Liljenroth for the ammonia reactor,

    Liljenroth, F. G., *Chem. Met. Eng.*, *19*, 287 (1918).

It was rediscovered by Van Heerden,

    Van Heerden, C., *Ind. Eng. Chem.*, *45*, 1242 (1953).

The data shown here are from

    Vejtasa, S. A., and R. A. Schmitz, *AIChE J.*, *16*, 410 (1970).

Identical phenomena in combustion reactions are discussed in detail in

    Frank-Kamenetskii, D. A., *Diffusion and Heat Transfer in Chemical Kinetics*, 2nd ed., Plenum, New York, 1969.

The data on an oscillatory reactor are from

> Baccaro, G. P., N. Y. Gaitonde, and J. M. Douglas, *AIChE J.*, *16*, 249 (1970).

Oscillations in a catalytic reaction are described in

> Beusch, H., P. Fieguth, and E. Wicke, in *Chemical Reaction Engineering, Advances in Chemistry Series No. 109*, American Chemical Society, Washington, D.C., 1972.

Further data and additional references may be found in review and contributed papers at the Fifth European/Second International Symposium on Chemical Reaction Engineering, published as

> *Chemical Reaction Engineering*, Elsevier, Amsterdam, 1972,

and in the review paper

> Denn, M. M., in V. W. Weekman, Jr., ed., *Annual Review of Industrial and Engineering Chemistry, 1970*, American Chemical Society, Washington, D.C., 1972.

Oscillations and structure in biological systems are summarized in

> Glansdorff, P., and I. Prigogine, *Thermodynamic Theory of Structure, Stability and Fluctuation*, Wiley, New York, 1971,

and in the research papers cited in the Postface.

The physical significance of the Reynolds number and data on the transition to turbulence in rectilinear flow are available in any book on fluid mechanics. See, for example,

> Bird, R. B., W. E. Stewart, and E. L. Lightfoot, *Transport Phenomena*, Wiley, New York, 1960.
>
> Schlichting, H., *Boundary-Layer Theory*, 6th ed., McGraw-Hill, New York, 1968.
>
> Whitaker, S., *Introduction to Fluid Mechanics*, Prentice-Hall, Englewood Cliffs, N.J., 1968.

Flow instabilities in molten polymers are reviewed in

> Tordella, J. P., in F. R. Eirich, ed., *Rheology*, vol. 5, Academic Press, New York, 1969.

Additional references are given in Chapter 12. For a summary of data on flow transition between rotating cylinders, see

> Chandrasekhar, S., *Hydrodynamic and Hydromagnetic Stability*, Oxford University Press, Inc., New York, 1961.

The data in Fig. 1.5 are from a University of Delaware M.Ch.E. Thesis by

J. J. Roisman. Transition data between rotating cylinders for a variety of non-Newtonian polymer solutions are presented in

Denn, M. M., and J. J. Roisman, *AIChE J.*, *15*, 454 (1969).

Data by several investigators on the onset of convective motion in the heat transfer experiments described in Sec. 1.4 are summarized in Chandrasekhar, cited above.

The mass transfer data in Fig. 1.9 are from

Olander, D. R., and L. B. Reddy, *Chem. Eng. Sci.*, *19*, 67 (1964).

Other anomalous mass transfer data associated with the surface-tension-induced instability are in

Berg, J. C., and G. S. Haselberger, *Chem. Eng. Sci.*, *26*, 481 (1971).
Clark, M. W., and C. J. King, *AIChE J.*, *16*, 64 (1970).
Linek, V., *Chem. Eng. Sci.*, 27, 627 (1972).

There is a nice simple experimental treatment of the Marangoni effect in

Ruckenstein, E., O. Smigelschi, and D. G. Sucin, *Chem. Eng. Sci.*, *25*, 1249 (1970).

Uniqueness in Lumped Parameter Systems

2

2.1 Introduction

In this first part of the book we shall examine the behavior of systems described by ordinary differential equations. The general form of such systems is

$$\dot{x}_1 \equiv \frac{dx_1}{dt} = f_1(x_1, x_2, x_3, \ldots, x_n, t)$$

$$\dot{x}_2 \equiv \frac{dx_2}{dt} = f_2(x_1, x_2, x_3, \ldots, x_n, t)$$

$$\vdots$$

$$\dot{x}_n \equiv \frac{dx_n}{dt} = f_n(x_1, x_2, x_3, \ldots, x_n, t)$$

(2-1a)

Here, $x_1, x_2, \ldots, x_n$ represent the n variables needed to define the *state* of the system at any time, t. $f_1, f_2, \ldots, f_n$ are n (usually nonlinear) functions of the state variables, $\{x_i\}$, and the independent variable, t. It will often be convenient to denote the collection $\{x_i\}$ as a *vector*, **x**,

16

$$\mathbf{x} = \begin{pmatrix} x_1 \\ x_2 \\ x_3 \\ \vdots \\ x_n \end{pmatrix}$$

in which case Eq. (2-1a) can be expressed compactly as

$$\dot{\mathbf{x}} \equiv \frac{d\mathbf{x}}{dt} = \mathbf{f}(\mathbf{x}, t) \tag{2-1b}$$

where $\mathbf{f}$ is a vector valued function of $\mathbf{x}$ and t. Equivalently, we may write Eqs. (2-1a) and (2-1b) in terms of a typical component,

$$\dot{x}_i \equiv \frac{dx_i}{dt} = f_i(\mathbf{x}, t), \qquad i = 1, 2, \ldots, n \tag{2-1c}$$

All three notational conventions are equivalent. We shall employ them interchangeably, using whichever is most convenient in a particular context.

There are small differences in certain notations for vector operations between control theory and classical nonlinear mechanics, on the one hand, and statistical and continuum mechanics on the other. We shall follow the latter. For example, $\mathbf{x} \cdot \mathbf{y}$ is the vector inner product, signifying

$$\mathbf{x} \cdot \mathbf{y} = \sum_{i=1}^{n} x_i y_i$$

Similarly, if $\mathbf{A}$ is a matrix, then $\mathbf{A} \cdot \mathbf{x}$ is a vector with a typical component

$$(\mathbf{A} \cdot \mathbf{x})_i = \sum_{j=1}^{n} A_{ij} x_j$$

On the other hand, a typical component of vector $\mathbf{x} \cdot \mathbf{A}$ is

$$(\mathbf{x} \cdot \mathbf{A})_j = \sum_{i=1}^{n} x_i A_{ij}$$

Similarly,

$$\mathbf{x} \cdot \mathbf{A} \cdot \mathbf{y} = \sum_{i=1}^{n} \sum_{j=1}^{n} x_i A_{ij} y_j$$

$$\mathbf{y} \cdot \mathbf{A} \cdot \mathbf{x} = \sum_{i=1}^{n} \sum_{j=1}^{n} y_i A_{ij} x_j = \sum_{i=1}^{n} \sum_{j=1}^{n} x_i A_{ji} y_j$$

Vector $\mathbf{x} \cdot \mathbf{A}$ is the same as $\mathbf{A}^T \cdot \mathbf{x}$, where $\mathbf{A}^T$ is the transpose of $\mathbf{A}$. In conventional matrix notation one of these would represent a row and the other a column vector, but the use of the dot product removes the need for that distinction.

2.2 Uniqueness of the Transient

Our analysis of dynamical systems depends on adherence to the *principle of causality*; namely, given the state of the system $\mathbf{x}(t_0) = \mathbf{x}_0$ at time t_0, the state $\mathbf{x}(t)$ at any later time t is uniquely defined by the solution to Eqs. (2-1) which passes through $\mathbf{x}_0$ at t_0. Causality is thus equivalent to uniqueness of the solution of the family of differential equations.

A sufficient condition for uniqueness can be readily derived. Suppose that $\mathbf{x}_1(t)$ and $\mathbf{x}_2(t)$ are both solutions to Eqs. (2-1) satisfying $\mathbf{x}(t_0) = \mathbf{x}_0$, and for simplicity take $t_0 = 0$. Assume that $\mathbf{x}_1(t)$ and $\mathbf{x}_2(t)$ both remain finite for $0 \le t < \infty$. Define

$$\mathbf{w}(t) = \mathbf{x}_1(t) - \mathbf{x}_2(t)$$

Then, from Eqs. (2-1),

$$\dot{\mathbf{w}}(t) = \mathbf{f}(\mathbf{x}_1(t), t) - \mathbf{f}(\mathbf{x}_2(t), t)$$

$$\mathbf{w}(t) = \int_0^t [\mathbf{f}(\mathbf{x}_1(\tau), \tau) - \mathbf{f}(\mathbf{x}_2(\tau), \tau)] \, d\tau \qquad (2\text{-}2)$$

The magnitude of a vector is defined as

$$|\mathbf{x}| = (\mathbf{x} \cdot \mathbf{x})^{1/2} = \left[\sum_{i=1}^{n} x_i^2 \right]^{1/2}$$

From Eq. (2-2), defining the magnitude of $\mathbf{w}(t)$ as $w(t)$,

$$|\mathbf{w}(t)| \equiv w(t) = \left| \int_0^t [\mathbf{f}(\mathbf{x}_1(\tau), \tau) - \mathbf{f}(\mathbf{x}_2(\tau), \tau)] \, d\tau \right|$$

and, using a standard inequality of the calculus,

$$w(t) \le \int_0^t |\mathbf{f}(\mathbf{x}_1(\tau), \tau) - \mathbf{f}(\mathbf{x}_2(\tau), \tau)| \, d\tau \qquad (2\text{-}3)$$

We now assume that $\mathbf{f}(\mathbf{x}, t)$ satisfies a *Lipschitz condition*:

$$|\mathbf{f}(\mathbf{x}_1, t) - \mathbf{f}(\mathbf{x}_2, t)| \le L_p(t) |\mathbf{x}_1 - \mathbf{x}_2| \qquad (2\text{-}4)$$

A sufficient condition for a continuously differentiable function is that all derivatives $\partial f_i/\partial x_j$ be bounded. Then Eq. (2-3) can be written as

$$w(t) \le \int_0^t L_p(\tau) w(\tau) \, d\tau \le L_{pm} \int_0^t w(\tau) \, d\tau \qquad (2\text{-}5)$$

where L_{pm} is the maximum value of $L_p(t)$ in the interval $(0, t)$. Now, let w_m be the maximum value of $w(t)$ in $(0, t)$. This maximum must exist, since $x_1(t)$ and $x_2(t)$ are bounded. Equation (2-5) then implies that

$$w(t) \le L_{pm} w_m t$$

in which case substitution directly back into Eq. (2-5) gives

$$w(t) \le L_{pm} \int_0^t w(\tau) \, d\tau \le L_{pm} \int_0^t L_{pm} w_m \tau \, d\tau = \frac{(L_{pm} t)^2}{2!} w_m$$

Substitution of this further result into Eq. (2-5) gives

$$w(t) \le L_{pm} \int_0^t w(\tau) \, d\tau \le L_{pm} \int_0^t \frac{(L_{pm}\tau)^2}{2!} w_m \, d\tau = \frac{(L_{pm} t)^3}{3!} w_m$$

After N substitutions we obtain

$$w(t) \le \frac{(L_{pm} t)^N}{N!} w_m \xrightarrow[N \to \infty]{} 0$$

Thus, a Lipschitz condition on $f(x, t)$ ensures uniqueness, since the difference between any two solutions must be zero.

2.3 Continuous Flow Reactor

The continuous flow stirred tank reactor (CFSTR) was introduced in Sec. 1.2 and shown schematically in Fig. 1.2. If we assume that the single liquid phase chemical reaction

$$A \longrightarrow \text{products}$$

is occurring in the reactor then the transient equations representing conservation of mass and energy can be written as

$$V \frac{dc}{d\tilde{t}} = q[c_f - c] - V r(c, T) \qquad (2\text{-}6)$$

$$\rho c_p V \frac{dT}{d\tilde{t}} = \rho c_p q[T_f - T] + [-\Delta H]Vr(c, T) - Ua[T - T_{cf}] \qquad (2\text{-}7)$$

Here, $\tilde{t}$ denotes time, so that t can be used subsequently for a dimensionless time variable. V and q are volume and volumetric flow rate, respectively; c is the concentration of reactant A; c_f the feed concentration; ρ and c_p the density and heat capacity, respectively, which are assumed to be constant and the same for feed and effluent; T the temperature; T_f the feed temperature; T_{cf} the coolant feed temperature; ΔH the heat of reaction; U an overall heat transfer coefficient; and a the area available for heat transfer. For illustrative purposes throughout the text the reaction will be assumed to be first order and irreversible, in which case the rate, $r(c, T)$, will have the form

$$r(c, T) = kce^{-E/RT} \qquad (2\text{-}8)$$

k and E are constants and R is the ideal gas constant.

The following dimensionless quantities are introduced:

$$x = \frac{c}{c_f} \qquad y = \frac{T}{T_f}$$

$$\alpha = \frac{kV}{q} \qquad \delta = \frac{Ua}{\rho q c_p} \qquad \beta = \frac{-\Delta H c_f}{\rho c_p T_f [1 + \delta]}$$

$$\gamma = \frac{E}{RT_f} \qquad t = \frac{q\tilde{t}}{V} \qquad \phi = \frac{1 + \delta T_{cf}/T_f}{1 + \delta}$$

It should be noted that β is positive for an exothermic (heat-liberating) reaction, for which $\Delta H < 0$. Equations (2-6) through (2-8) can then be written as

$$\frac{dx}{dt} = 1 - x - \alpha x e^{-\gamma/y} \qquad (2\text{-}9)$$

$$\frac{1}{1 + \delta} \frac{dy}{dt} = \phi - y + \alpha \beta x e^{-\gamma/y} \qquad (2\text{-}10)$$

For an adiabatic reactor, in which the heat transfer area goes to zero,

$$\text{adiabatic:} \quad \delta \to 0, \qquad \phi \to 1 \qquad (2\text{-}11)$$

Equations (2-9) and (2-10) will represent a primary example system in the text. It is readily established that the system satisfies a Lipschitz condition, and hence the transient solution for a given initial concentration-temperature pair, x_0 and y_0, will be unique.

2.4 Steady State

The *steady state*, or *equilibrium state*, of a dynamical system occurs when there are no changes in time, so time derivatives vanish. For Eqs. (2-1) a steady state will usually occur only for *autonomous systems*, in which $\mathbf{f}$ does not depend explicitly on t:

$$\dot{\mathbf{x}} = \mathbf{f}(\mathbf{x}) \tag{2-12}$$

Then the steady state, $\mathbf{x}_s$, is the solution of the algebraic equations

$$\mathbf{0} = \mathbf{f}(\mathbf{x}_s) \tag{2-13}$$

Nonlinear equations will usually have multiple solutions. For example, the equation

$$\dot{x} = -x[1 - x]$$

which is a simple model for batch bacterial growth, has steady states at solutions $x_s = 0, 1$ of

$$0 = -x_s[1 - x_s]$$

A sufficient condition for uniqueness of the steady state can be easily derived. Equation (2-13) is rewritten as

$$\mathbf{x}_s = \mathscr{F}(\mathbf{x}_s) \tag{2-14}$$

If there are two solutions, $\mathbf{x}_{s1}$ and $\mathbf{x}_{s2}$, then the difference between them is

$$\mathbf{w} = \mathbf{x}_{s1} - \mathbf{x}_{s2} = \mathscr{F}(\mathbf{x}_{s1}) - \mathscr{F}(\mathbf{x}_{s2}) \tag{2-15}$$

Let $|\mathbf{w}| \equiv w$ and assume that $\mathscr{F}(\mathbf{x})$ satisfies a Lipschitz condition. Then Eq. (2-15) becomes

$$w = |\mathscr{F}(\mathbf{x}_{s1}) - \mathscr{F}(\mathbf{x}_{s2})| \le L_p w \tag{2-16}$$

Equation (2-16) can have no solution except $w = 0$ if $L_p < 1$. This will usually be a conservative condition, as we shall see in the next section.

2.5 Steady-State CFSTR

The steady-state description of the continuous flow stirred tank reactor is obtained by setting time derivatives to zero in Eqs. (2-9) and (2-10):

$$0 = 1 - x_s - \alpha x_s e^{-\gamma/y_s} \tag{2-17}$$

$$0 = \phi - y_s + \alpha \beta x_s e^{-\gamma/y_s} \tag{2-18}$$

A simple relation between the steady-state concentration and temperature is obtained by multiplying Eq. (2-17) by β and adding to Eq. (2-18):

$$\beta x_s = \beta + \phi - y_s \tag{2-19}$$

Since the concentration can never exceed the feed concentration at steady state or fall below zero, we have

$$0 \le x_s \le 1 \tag{2-20a}$$

in which case Eq. (2-19) implies that

$$\phi \le y_s \le \phi + \beta \tag{2-20b}$$

It is helpful to combine Eqs. (2-18) and (2-19) to obtain a single equation for temperature, which can be written as

$$\frac{1}{\alpha}[y_s - \phi] = F(y_s) \tag{2-21}$$

$$F(y) \equiv [\beta + \phi - y]e^{-\gamma/y} \tag{2-22}$$

The function $F(y)$ has the general shape shown in Fig. 2.1, with an inflection at y_i and a maximum at y_m. The first two derivatives are

$$F'(y) = \{\gamma[\phi + \beta - y] - y^2\}y^{-2}e^{-\gamma/y} \tag{2-23}$$

$$F''(y) = \gamma\{\gamma[\phi + \beta - y] - 2[\phi + \beta]y\}y^{-4}e^{-\gamma/y} \tag{2-24}$$

y_i and y_m occur, respectively, at the vanishing of $F''(y)$ and $F'(y)$:

$$y_i = \frac{\gamma[\phi + \beta]}{\gamma + 2[\phi + \beta]} \tag{2-25a}$$

$$y_m = -\frac{\gamma}{2} + \frac{1}{2}\{\gamma^2 + 4\gamma[\phi + \beta]\}^{1/2} \tag{2-25b}$$

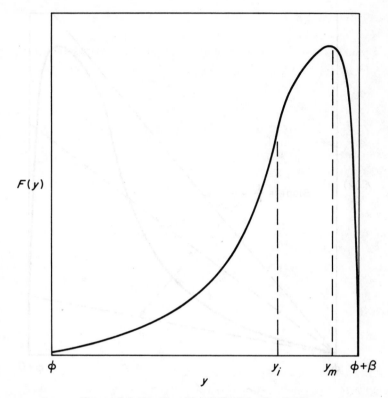

Figure 2.1 Function $F(y)$, defined by Eq. (2-22).

Figure 2.2 shows the line $[y - \phi]/\alpha$ plotted for various α on the same coordinates with the function $F(y)$. According to Eq. (2-21), the steady state will occur when the two lines intersect. It is clear by inspection of the figure that there is a range of α for which *three* intersections will occur, and hence there will be three steady-state solutions to Eqs. (2-17) and (2-18). α is proportional to the residence time, $\theta = V/q$, so there will be a range of residence times for which three steady states are predicted. The solid line in Fig. 1.3 represents the temperature-residence time solution for the reaction system studied by Vejtasa and Schmitz. In the range of multiple solutions they observed two of the three theoretically possible states.

The results of the preceding section can be applied to the reactor system to define a range of parameters for which a unique solution will occur. Equation (2-21) is rewritten as

$$y_s = \phi + \alpha F(y_s) \equiv \mathscr{F}(y_s)$$

which has the form of Eq. (2-14). The Lipschitz constant is

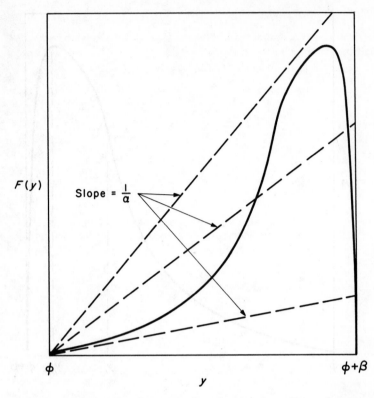

Figure 2.2 Possibility of one or three steady states, depending on slope $1/\alpha$.

$$L_p = \max_{\phi \le y \le \phi+\beta} |\mathscr{F}'(y)| = \alpha \max_{\phi \le y \le \phi+\beta} |F'(y)|$$

Both positive and negative slopes need to be considered, leading to the following possibilities:

$$\gamma > \frac{2\phi[\phi+\beta]}{\beta} \qquad L_p = \max \begin{cases} \alpha e^{-\gamma/[\phi+\beta]} \\ \alpha\left\{\frac{2}{\gamma}[\gamma + 2\{\phi+\beta\}]^2 - 1\right\}e^{-2}e^{-\gamma/[\phi+\beta]} \end{cases}$$

$$\frac{2\phi[\phi+\beta]}{\beta} > \gamma > \frac{\phi^2}{\beta} \qquad L_p = \max \begin{cases} \alpha e^{-\gamma/[\phi+\beta]} \\ \alpha\left\{\frac{\beta\gamma}{\phi^2} - 1\right\}e^{-\gamma/\phi} \end{cases}$$

In each case the upper value corresponds to the region of negative slope, and the lower to the largest positive slope. The sufficient condition for uniqueness is $L_p < 1$.

It is evident from Fig. 2.2 that only the region in which $F(y)$ has a positive slope is relevant to uniqueness. This can be established as follows, where we exploit the fact that we are dealing with a single scalar equation. Let y_{1s} and y_{2s} be two solutions to Eq. (2-21). Then

$$w = y_{1s} - y_{2s} = \alpha[F(y_{1s}) - F(y_{2s})]$$

or, applying the mean value theorem to the continuously differentiable function $F(y)$,

$$w = \alpha F'(\bar{y})w \qquad (2\text{-}26)$$

$\bar{y}$ is a point between y_{1s} and y_{2s}. There can be no solution to Eq. (2-26) if $F'(\bar{y})$ is negative. We shall return to this point subsequently. There is no loss in generality in defining $y_{1s} > y_{2s}$, so $w > 0$. Since $F'(\bar{y}) \leq \max_{\phi \leq y \leq \phi+\beta} F'(y)$ we have, from Eq. (2-26),

$$w \leq \alpha \left[\max_{\phi \leq y \leq \phi+\beta} F'(y) \right] w \qquad (2\text{-}27)$$

in which case uniqueness is ensured for

$$\alpha \left[\max_{\phi \leq y \leq \phi+\beta} F'(y) \right] < 1 \qquad (2\text{-}28)$$

We have noted that uniqueness is ensured when $F'(\bar{y}) < 0$. This is ensured if $\max F'(y) < 0$, or, from Eq. (2-23),

$$\max_{\phi \leq y \leq \phi+\beta} \gamma[\phi + \beta - y] - y^2 < 0$$

The function decreases with y, having a maximum at $y = \phi$. Thus, we obtain the condition for uniqueness

$$\beta\gamma < \phi^2 \qquad (2\text{-}29)$$

This result is independent of α, but it is quite conservative, since it requires that the maximum of $F(y)$ occur at $y = \phi$.

A stronger uniqueness condition independent of α is obtained by noting that Eq. (2-21) can be written as

$$\frac{F(y_s)}{y_s - \phi} = \frac{1}{\alpha} \qquad (2\text{-}30)$$

The left-hand side of Eq. (2-30) varies from infinity as $y_s \to \phi$ to zero as $y_s \to \phi + \beta$. If the left side is monotonically decreasing in the region $\phi \leq y_s \leq \phi + \beta$, there can be only one value of y_s for which the left side takes on the value $1/\alpha$. Thus, a sufficient condition for uniqueness is

$$\frac{d}{dy} \frac{F(y)}{y - \phi} \leq 0 \quad \text{in} \quad \phi \leq y \leq \phi + \beta$$

or

$$\max_{\phi \leq y \leq \phi + \beta} [y - a]F'(y) - F(y) \leq 0$$

The maximum occurs at $y = y_i$, Eq. (2-25a), giving the result

$$\beta \gamma < 4\phi^2 + 4\beta\phi \qquad (2\text{-}31)$$

Equation (2-31) is a considerably stronger uniqueness condition than Eq. (2-20).

The fact that the middle steady state in Fig. 1.3 was not observed experimentally can be explained qualitatively by reference to Fig. 2.2. Suppose that the system is disturbed slightly from the steady state such that Eq. (2-19) is still satisfied:

$$\beta x = \beta + \phi - y$$

Then the temperature equation, (2-10), will become

$$\frac{1}{1 + \delta} \frac{dy}{dt} = \phi - y + \alpha F(y) \qquad (2\text{-}32)$$

It is convenient to refer temperatures to the steady state,

$$y = y_s + \eta$$

in which case Eq. (2-32) can be written as

$$\frac{1}{1 + \delta} \frac{d\eta}{dt} = \phi - y_s - \eta + \alpha F(y_s + \eta) \qquad (2\text{-}33)$$

Here, we have used the fact that $dy_s/dt = 0$. From the mean value theorem, $F(y_s + \eta) = F(y_s) + F'(\bar{y})\eta$, where $\bar{y}$ lies between y_s and $y_s + \eta$. Together with Eqs. (2-21) and (2-33) we thus obtain

$$\frac{1}{1+\delta}\frac{d\ln\eta}{dt} = -1 + \alpha F'(\bar{y}) \qquad (2\text{-}34)$$

The steady state will be unstable if $\ln\eta$ grows with time, since in that case the disturbed system will move even farther away from y_s with time. It will therefore be impossible to maintain the steady state in the face of small disturbances. According to Eq. (2-34), this will occur when $F'(y) > 1/\alpha$. That is, the steady state will be unstable when, in the neighborhood of the steady-state temperature, the slope of $F(y)$ is greater than the slope of the straight line $[y - \phi]/\alpha$. From Fig. 2.2 this condition applies to the middle steady state.

This *slope condition* is sometimes interpreted thermodynamically by noting that, at steady state, $F(y)$ represents the rate of heat input to the system because of chemical reaction, while $\phi - y$ represents the net rate of heat removal by flow. If the rate of heat input exceeds the rate of removal following a small temperature increase, as will occur at the middle steady state, the temperature will continue to grow even more, causing instability. This argument is a useful one, but it must be applied very cautiously since the thermodynamic interpretations apply only at steady state, while we are using them in a nonsteady situation.

The converse of the slope condition cannot be applied at the high and low steady states. We have assumed a very special disturbance with a specific relation between x and y. If the system is unstable in the face of that disturbance, as at the middle steady state, then we can say with certainty that the state cannot be maintained. If it is stable with respect to that disturbance, however, we have proved nothing, for the state might be unstable to some other disturbance.

2.6 Anisotropic Fluid

In an incompressible Newtonian liquid the stress at any point is directly proportional to the deformation rate, with the constant viscosity as the proportionality factor. In Cartesian coordinates this relation is written as

$$\tau_{ij} + p\,\delta_{ij} = 2\mu\,d_{ij}, \qquad i,j = 1,2,3 \qquad (2\text{-}35)$$

$$d_{ij} = \frac{1}{2}\left[\frac{\partial v_i}{\partial x_j} + \frac{\partial v_j}{\partial x_i}\right] \qquad (2\text{-}36)$$

$$\delta_{ij} = \begin{cases} 1, & i = j \\ 0, & i \neq j \end{cases}$$

p is the isotropic pressure, $\mathbf{v}$ the velocity vector, $\mathbf{x}$ the coordinate location, and $\boldsymbol{\tau}$ the extra-stress tensor. μ is the viscosity. In a simple shear flow, defined by (Fig. 2.3)

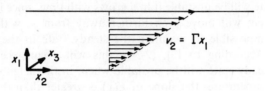

Figure 2.3 Simple shear flow.

$$v_1 = v_3 = 0 \qquad v_2 = \Gamma x_1 \tag{2-37}$$

the only nonzero stresses are the shear stresses, which follow the relation

$$\tau_{12} = \tau_{21} = \mu \Gamma \tag{2-38}$$

The viscosity is then measured as the ratio of shear stress to shear rate, Γ.

Polymer solutions and melts, fiber suspensions, and liquid crystals, because of the internal structure in the fluid, do not follow this simple relation. A class of structural theories has been developed which relate the stress to the local structure. In the simplest of these theories, developed by J. L. Ericksen, the structure in the liquid is described by the magnitude and orientation of a vector, $\mathbf{n}$. The stress-deformation rate relation in this theory is

$$\tau_{ij} + p\,\delta_{ij} = \left[\mu_1 \sum_{r,s=1}^{3} d_{rs} n_r n_s\right] n_i n_j + 2\mu_2\,d_{ij} + 2\mu_3 \sum_{k=1}^{3} [d_{ik} n_k n_j + n_i\,d_{jk} n_k] \tag{2-39}$$

$$\frac{\partial n_i}{\partial t} + \sum_{k=1}^{3} v_k \frac{\partial n_i}{\partial x_k} - \frac{1}{2} \sum_{k=1}^{3} \left[\frac{\partial v_i}{\partial x_k} - \frac{\partial v_k}{\partial x_i}\right] n_k$$
$$= \left[\beta_1 + \beta_2 \sum_{r,s=1}^{3} d_{rs} n_r n_s\right] n_i + \beta_3 \sum_{k=1}^{3} d_{ik} n_k \tag{2-40}$$

In a simple shear flow defined by Eq. (2-37) the shear stress and orientation vector equations simplify to

$$\tau_{12} = \{\mu_1 n_1^2 n_2^2 + \mu_2 + \mu_3[n_1^2 + n_2^2]\}\Gamma \tag{2-41}$$

$$\frac{dn_1}{dt} = [\beta_1 + \beta_2 \Gamma n_1 n_2]n_1 + \frac{1}{2}[\beta_3 - 1]\Gamma n_2 \tag{2-42a}$$

$$\frac{dn_2}{dt} = [\beta_1 + \beta_2 \Gamma n_1 n_2]n_2 + \frac{1}{2}[\beta_3 + 1]\Gamma n_1 \qquad (2\text{-}42\text{b})$$

$$\frac{dn_3}{dt} = [\beta_1 + \beta_2 \Gamma n_1 n_2]n_3 \qquad (2\text{-}42\text{c})$$

It is assumed that $\mathbf{n}$ is spatially independent. In the general theory the co-efficients $\mu_1, \mu_2, \mu_3, \beta_1, \beta_2, \beta_3$ may be functions of the magnitude of $\mathbf{n}$. For simplicity we shall assume that they are constant, which is in accord with most flow calculations using the theory.

From our point of view, the interesting feature of this theory is the possibility of multiple solutions. Clearly Eqs. (2-42) admit the steady-state solution

$$\mathbf{n}_s = 0$$

In that case the shear stress in Eq. (2-41) is proportional to the shear rate, the theory degenerates to the Newtonian fluid, and μ_2 is the Newtonian viscosity. There are, however, other steady states, with $n_3 = 0$ and n_1 and n_2 solutions of

$$0 = [\beta_1 + \beta_2 \Gamma n_1 n_2]n_1 + \tfrac{1}{2}[\beta_3 - 1]\Gamma n_2 \qquad (2\text{-}43\text{a})$$

$$0 = [\beta_1 + \beta_2 \Gamma n_1 n_2]n_2 + \tfrac{1}{2}[\beta_3 + 1]\Gamma n_1 \qquad (2\text{-}43\text{b})$$

It is useful to note that Eq. (2-42c) can be written as

$$\frac{d \ln n_3}{dt} = \beta_1 + \beta_2 \Gamma n_1 n_2$$

If n_3 deviates slightly from its equilibrium state of $n_3 = 0$, it will continue to grow unless

$$\beta_1 + \beta_2 \Gamma n_1 n_2 < 0 \qquad (2\text{-}44)$$

This inequality is therefore a restriction on n_1 and n_2 for a stable steady state. Equations (2-43a) and (2-43b) combine to give the following relations:

$$0 = [\beta_1 + \beta_2 \Gamma n_1 n_2]n^2 + \beta_3 \Gamma n_1 n_2 \qquad (2\text{-}45)$$

$$n_1^2 = \frac{\beta_3 - 1}{2\beta_3} n^2 \qquad n_2^2 = \frac{\beta_3 + 1}{2\beta_3} n^2 \qquad (2\text{-}46)$$

$$n^2 = n_1^2 + n_2^2$$

Equation (2-44), combined with Eq. (2-45), implies that

$$\beta_3 \Gamma n_1 n_2 > 0 \qquad (2\text{-}47)$$

Equations (2-46) can be written as

$$2\beta_3 \Gamma n_1 n_2 = \pm[\beta_3^2 - 1]^{1/2} \Gamma n^2 > 0 \qquad (2\text{-}48)$$

where the inequality follows from Eq. (2-47). Thus, one of the two roots represents an unstable state. Since Γ can be positive or negative, the possible stable steady state is governed by the relation

$$\Gamma n_1 n_2 = \frac{[\beta_3^2 - 1]^{1/2}}{2\beta_3} n^2 |\Gamma| \qquad (2\text{-}49)$$

n^2 is then obtained from Eq. (2-45) as

$$n^2 = -\beta_3 \frac{2\beta_1 + [\beta_3^2 - 1]^{1/2} |\Gamma|}{\beta_2 [\beta_3^2 - 1]^{1/2} |\Gamma|} \qquad (2\text{-}50)$$

It then follows from Eq. (2-41) that the apparent viscosity, τ_{12}/Γ, is a function of shear rate, with the functionality defined by Eqs. (2-49) and (2-50). Note that a real steady-state solution cannot exist if $-1 < \beta_3 < +1$.

According to this simple anisotropic fluid model, there are three possible steady-state stress distributions in simple shearing flow. One cannot be stable, one gives a shear dependent viscosity, and in one the orientation vector vanishes and Newtonian behavior is recovered. We shall show subsequently that there is a critical shear rate,

$$\Gamma_c^2 = \frac{4\beta_1^2}{\beta_3^2 - 1}$$

Below this value the solution $\mathbf{n} = \mathbf{0}$ is stable and the anisotropic fluid will exhibit Newtonian behavior. Above this value the Newtonian solution is not stable, and the stable configuration is the one which exhibits shear dependent properties. This behavior is qualitatively like that observed in polymer solutions, in which the viscosity takes on a constant *zero-shear* value at low shear rates, followed by a region in which viscosity decreases with shear rate, followed finally by a third *upper-Newtonian* region in which viscosity is again constant. Note that for large $|\Gamma|$, $\mathbf{n}$ goes to a constant, nonzero value which is independent of shear rate.

2.7 Concluding Remarks

In physical systems described by nonlinear differential equations it is evident that multiple steady-state solutions will often occur and that the equilibrium state actually observed will depend on the stability of the states. The two examples considered in this chapter share that common feature, but they are otherwise representative of the very wide scope of physical situations to which these observations apply. We shall return to the examples of this chapter when we develop techniques for examining the stability of a steady state. We shall return to the question of multiple steady states subsequently in the discussion of distributed parameter systems.

BIBLIOGRAPHICAL NOTES

The results on uniqueness can be found in any good book on ordinary differential equations, such as

> Bellman, R., *Stability Theory of Differential Equations*, McGraw-Hill, New York, 1953.
>
> Coddington, E. A., and N. Levinson, *Theory of Ordinary Differential Equations*, McGraw-Hill, New York, 1955.

For a derivation of the continous flow reactor equations, see

> Aris, R., *Introduction to the Analysis of Chemical Reactors*, Prentice-Hall, Englewood Cliffs, N.J., 1965.
>
> Russell, T. W. F., and M. M. Denn, *Introduction to Chemical Engineering Analysis*, Wiley, New York, 1972.

The nomenclature is patterned after that of Aris in a survey paper,

> Aris, R., *Chem. Eng. Sci.*, 24, 149 (1969).

The uniqueness results given in the chapter are in the paper by Aris and in

> Perlmutter, D. D., *Stability of Chemical Reactors*, Prentice-Hall, Englewood Cliffs, N.J., 1972.

The development leading to the uniqueness criterion (2-31) is from

> Luss, D., *Chem. Eng. Sci.*, 26, 1713 (1971).

The model of an anisotropic fluid is by Ericksen,

> Ericksen, J. L., *Trans. Soc. Rheol.*, 4, 29 (1960); 6, 275 (1962).

For some more recent treatments and comparison with data on polymer solutions, see

> Denn, M. M., and A. B. Metzner, *Trans. Soc. Rheol.*, 10, 215 (1966).
>
> Gordon, R. J., and W. R. Schowalter, *Trans. Soc. Rheol.*, 16, 79 (1972).
>
> Ling, C. C., S. J. Allen, and K. A. Kline, *Trans. Soc. Rheol.*, 16, 129 (1972).

Stability   **3**

3.1 Introduction

In developing procedures for studying the stability of an equilibrium state it is helpful to use some geometrical ideas. We shall restrict attention here to systems described by autonomous ordinary differential equations of the form of Eq. (2-12),

$$\dot{\mathbf{x}} = \mathbf{f}(\mathbf{x}) \tag{3-1}$$

The steady state satisfies

$$0 = \mathbf{f}(\mathbf{x}_s) \tag{3-2}$$

$\mathbf{x}(t)$ can be looked upon as a point in an n-dimensional Euclidean space, as shown pictorially in Fig. 3.1 for two dimensions. This space is often called the *phase space*. As time progresses, the solution to Eq. (3-1) will move through that space. If we enclose a region of the phase space by a surface,

$$\mathscr{S}(\mathbf{x}) = 0 \tag{3-3}$$

then we will say that *the system* (3-1) *is stable with respect to the region* (3-3) *if* $\mathbf{x}(t)$ *remains within the region enclosed by* $\mathscr{S}(\mathbf{x}) = 0$ *for all time* $0 \leq t < \infty$.

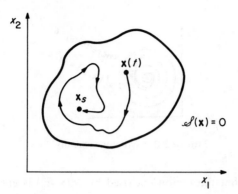

Figure 3.1 Movement of a point $\mathbf{x}(t)$ in a region of phase space bounded by a closed surface $\mathscr{S}(\mathbf{x}) = 0$.

Note that this definition of stability is less stringent than the one which we have been using informally thus far. A system is considered stable with respect to a region so long as the transient never leaves that region, even though the system may never return to the steady state. Note, too, that the choice of the region is an essential part of the stability consideration. As demonstrated in Fig. 3.2, a system whose transient returns to the steady state following a disturbance might still be unstable for a given choice of $\mathscr{S}(\mathbf{x})$. Since the size of the region in which fluctuations are tolerated will often be governed by practical engineering considerations, stability with respect to a region is often called *practical stability*.

The stronger notion of stability, which requires that following a disturbance the system must return to its steady state, can be incorporated into the concept developed here by considering a one-parameter family of surfaces,

$$\mathscr{S}(\mathbf{x}, c) = 0 \tag{3-4}$$

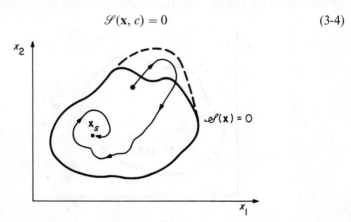

Figure 3.2 A trajectory which is unstable with respect to one region and stable with respect to another.

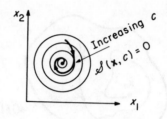

Figure 3.3 Asymptotic stability.

$0 \leq c < \infty$, such that the region enclosed by $\mathscr{S}(\mathbf{x}, c_n)$ is entirely contained within the region enclosed by $\mathscr{S}(\mathbf{x}, c_m)$, $c_n < c_m$, and the steady state, $\mathbf{x}_s$, is contained within all regions in the family. (That is, as $c \to 0$ the region shrinks to the point $\mathbf{x}_s$.) Then we say that *the system* (3-1) *is asymptotically stable with respect to the family of regions* (3-4) *if when* $\mathbf{x}(t)$ *lies in a region enclosed by* $\mathscr{S}(\mathbf{x}, c_m) = 0$, *then* $\mathbf{x}(t + \Delta)$, $\Delta > 0$, *lies in a region enclosed by* $\mathscr{S}(\mathbf{x}, c_n) = 0$, $c_n < c_m$. This concept of asymptotic stability is illustrated pictorially in Fig. 3.3.

3.2 Basic Inequality

We can develop quantitative descriptions for the definitions introduced in the preceding section by continuing to think geometrically. Let $\mathbf{n}$ be the outward-directed normal on the surface $\mathscr{S}(\mathbf{x}) = 0$, as shown in Fig. 3.4. For the dynamical system (3-1) the vector $\mathbf{f}(\mathbf{x})$ represents the rate of change of position in the phase space, or the velocity. If, when $\mathbf{x}(t)$ lies on the surface, the velocity vector $\mathbf{f}$ points out, the system will leave the region. Similarly, if

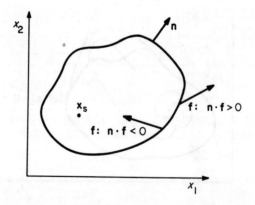

Figure 3.4 A region in phase space bounded by a surface $\mathscr{S}(\mathbf{x}) = 0$. $\mathbf{n}$ is the outward normal and $\mathbf{f}$ the velocity of the point $\mathbf{x}$.

$\mathbf{f}$ points in or along the surface, the system will remain within the region. The vector $\mathbf{f}$ points into the region if the projection of $\mathbf{f}$ on the outward normal is negative, and $\mathbf{f}$ points along the surface (tangent) if the projection of $\mathbf{f}$ on $\mathbf{n}$ is zero. Thus, *a necessary and sufficient condition for stability of the system* (3-1) *with the respect to the region* (3-3) *is*

$$\mathbf{n} \cdot \mathbf{f}(\mathbf{x}) \leq 0 \qquad \text{everywhere on } \mathscr{S}(\mathbf{x}) = 0 \qquad (3\text{-}5)$$

If there are points on the surface where an outward normal is not uniquely defined, then inequality (3-5) must be satisfied in the limit from all directions. Note that, within an arbitrary positive scaling factor,

$$\mathbf{n} = \nabla \mathscr{S} \qquad (3\text{-}6)$$

where $\nabla \mathscr{S}$ (gradient of $\mathscr{S}$) is the vector with components $\partial \mathscr{S}/\partial x_i$.

According to the definition for asymptotic stability, as time progresses the system must move steadily through the family of surfaces (3-4) with continuously decreasing c. Clearly, this requires that $\mathbf{x}$ always move into the region from the surface. Thus, *a necessary and sufficient condition for asymptotic stability of system* (3-1) *with respect to the family of regions* (3-4) *is*

$$\mathbf{n} \cdot \mathbf{f} < 0 \qquad \text{everywhere on } \mathscr{S}(\mathbf{x}, c) = 0, \qquad c < c_\infty \qquad (3\text{-}7)$$

The value c_∞ defines the limiting region of asymptotic stability.

3.3 Asymptotic Stability

We shall usually be concerned with the behavior of nonlinear processes. Very few nonlinear differential equations can be treated analytically, however; but some very useful information can be obtained by restricting attention to the region very close to the steady state and considering only the effect of infinitesimal disturbances.

It is convenient to change variables so that the frame of reference is in terms of the steady state. Define

$$\boldsymbol{\xi} = \mathbf{x} - \mathbf{x}_s \qquad (3\text{-}8)$$

The steady state is then at $\boldsymbol{\xi} = \mathbf{0}$. Since $\mathbf{x}_s$ is a constant, $\dot{\boldsymbol{\xi}} = \dot{\mathbf{x}}$, and Eq. (3-1) can be written as

$$\dot{\boldsymbol{\xi}} = \mathbf{f}(\mathbf{x}_s + \boldsymbol{\xi}) \qquad (3\text{-}9)$$

If **f** is twice differentiable at $\mathbf{x} = \mathbf{x}_s$ with bounded second derivatives, then we may use Taylor's theorem to write

$$\mathbf{f}(\mathbf{x}_s + \boldsymbol{\xi}) = \mathbf{f}(\mathbf{x}_s) + \mathbf{A} \cdot \boldsymbol{\xi} + \mathbf{o}(\boldsymbol{\xi}) \tag{3-10}$$

From Eq. (3-2), $\mathbf{f}(\mathbf{x}_s) = \mathbf{0}$. The matrix **A** is the gradient of **f** evaluated at the steady state; that is, **A** is a constant matrix.

$$\mathbf{A} = \nabla \mathbf{f}(\mathbf{x}_s) \qquad \nabla \circ f(x_s) \, ?$$

$$A_{ij} = \frac{\partial f_i(x_s)}{\partial x_j} \tag{3-11}$$

$\mathbf{o}(\boldsymbol{\xi})$ represents terms which go to zero faster than $|\boldsymbol{\xi}|$:

$$\lim_{|\boldsymbol{\xi}| \to 0} \frac{\mathbf{o}(\boldsymbol{\xi})}{|\boldsymbol{\xi}|} = 0 \tag{3-12}$$

Equation (3-9) can then be written as $\qquad f(x_s) = 0$

$$\dot{\boldsymbol{\xi}} = \mathbf{A} \cdot \boldsymbol{\xi} + \mathbf{o}(\boldsymbol{\xi}) \tag{3-13}$$

We shall study the asymptotic stability of this system.

For the family of surfaces (3-4) we shall use ellipses of mean radius r:

$$\mathscr{S}(\boldsymbol{\xi}, r) = \tfrac{1}{2} \boldsymbol{\xi} \cdot \mathbf{B} \cdot \boldsymbol{\xi} - \tfrac{1}{2} r^2 = 0 \tag{3-14}$$

The matrix **B** must be positive definite in order to obtain closed elliptical contours. **B** can be taken as symmetric ($\mathbf{B}^T = \mathbf{B}$) without loss of generality, since only the symmetric part of a matrix enters into the terms of a quadratic form. The factor of $\tfrac{1}{2}$ is for convenience in differentiation. The outward normal, Eq. (3-6), is then the gradient of $\mathscr{S}$, which is simply

$$\mathbf{n} = \boldsymbol{\xi} \cdot \mathbf{B}$$

and inequality (3-7) is formed by taking the inner product of the right-hand side of Eq. (3-13) with $\boldsymbol{\xi} \cdot \mathbf{B}$:

$$\boldsymbol{\xi} \cdot \mathbf{B} \cdot \mathbf{A} \cdot \boldsymbol{\xi} + o(r^2) < 0 \tag{3-15}$$

Note that $\boldsymbol{\xi} \cdot \mathbf{o}(\boldsymbol{\xi}) = o(|\boldsymbol{\xi}|^2) = o(r^2)$, since only orders of magnitude are involved.

One further change of variables is helpful to normalize the system,

$$\mathbf{y} = \frac{\boldsymbol{\xi}}{r}$$

$$|\mathbf{y}| = \frac{\sqrt{\boldsymbol{\xi} \cdot \boldsymbol{\xi}}}{r} = 1 \tag{3-16}$$

Dividing inequality (3-15) by r^2 we can write

$$\mathbf{y} \cdot \mathbf{B} \cdot \mathbf{A} \cdot \mathbf{y} + \frac{o(r^2)}{r^2} < 0 \tag{3-17}$$

For infinitesimal perturbations away from the steady state we let $r^2 \to 0$. In that case, using the definition (3-12) for $o(r^2)$, we obtain, finally,

$$\mathbf{y} \cdot \mathbf{B} \cdot \mathbf{A} \cdot \mathbf{y} < 0 \tag{3-18}$$

That is, the necessary and sufficient condition for asymptotic stability with respect to a family of elliptical surfaces is that the matrix $\mathbf{B} \cdot \mathbf{A}$ be negative definite, where $\mathbf{B}$ is positive definite and $\mathbf{A} = \nabla \mathbf{f}(\mathbf{x}_s)$. As shown in Appendix A, this is equivalent to requiring that *all eigenvalues of the matrix* $\mathbf{A}$ *must have negative real parts.*

We shall apply this result extensively in the next chapter, so some further comments are in order. First, it is necessary to study asymptotic stability here so that we have a strong inequality (<0) in Eqs. (3-17) and (3-18). If we had weak inequality (≤ 0), then Eq. (3-18) would admit eigenvalues with zero real parts. In that case we could always find a disturbance $\mathbf{y}$ such that $\mathbf{y} \cdot \mathbf{B} \cdot \mathbf{A} \cdot \mathbf{y}$ vanished and the $o(r^2)$ term was the only one left in Eq. (3-17). Thus, *when there are eigenvalues of the matrix* $\mathbf{A}$ *with zero real parts we cannot draw any conclusion about stability of the nonlinear equation to infinitesimal disturbances.*

Second, a bit of reflection leads to the conclusion that if any eigenvalue of $\mathbf{A}$ has a positive real part, then there will always be at least one disturbance such that

$$\mathbf{y} \cdot \mathbf{B} \cdot \mathbf{A} \cdot \mathbf{y} + \frac{o(r^2)}{r^2} > 0$$

or, equivalently,

$$\mathbf{n} \cdot \mathbf{f} > 0$$

Furthermore, if **f** points outward from all members of a family of elliptical regions, it will point outward from any family of simply connected smooth surfaces. Thus, *it is sufficient for instability to infinitesimal disturbances with respect to any region for an eigenvalue of* **A** *to have a positive real part.* The result obtained using an elliptical family of regions therefore has widespread generality.

3.4 Linear Homogeneous Systems

The result of the preceding section on asymptotic stability of nonlinear systems suggests direct comparison with the linear homogeneous equation,

$$\dot{\xi} = \mathbf{A} \cdot \xi \tag{3-19}$$

The only steady state is at $\xi = \mathbf{0}$. The solution will have the form

$$\xi(t) = \sum_k \mathbf{c}_k e^{\lambda_k t} \tag{3-20}$$

where the $\mathbf{c}_k$ are constant eigenvectors and λ_k the eigenvalues of **A**.* Clearly, if the real parts of all λ_k are negative, then $\xi(t)$ goes to zero as t approaches infinity, and the system is asymptotically stable. If even one eigenvalue has a positive real part, there will be a term with exponentially growing magnitude, $|\xi(t)|$ will increase without bound as t goes to infinity, and the system is unstable. Until $|\xi(t)|$ grows large this is the behavior obtained for the nonlinear system in the preceding section. Thus, we can conclude that *for infinitesimal perturbations the behavior of the nonlinear system* (3-1) *is described by the linear system* (3-19) *with* $\mathbf{A} = \nabla \mathbf{f}(\mathbf{x}_s)$ *as long as no eigenvalue of* **A** *has a zero real part.* Use of the linearized equations to study behavior of a nonlinear system is sometimes called *Liapunov's first method.*

3.5 Higher-Order Systems

Physical systems may sometimes be described by differential equations with second and higher time derivatives. These can always be reduced to the form

*

$$\dot{\xi} = \sum_k \lambda_k \mathbf{c}_k e^{\lambda_k t} = \mathbf{A} \cdot \xi = \sum_k \mathbf{A} \cdot \mathbf{c}_k e^{\lambda_k t}$$

Then

$$\mathbf{A} \cdot \mathbf{c}_k = \lambda_k \mathbf{c}_k$$

which is the eigenvalue equation.

of Eq. (3-1). Consider, for example, the equation

$$\ddot{x} = g(x, \dot{x}) \tag{3-21}$$

Define

$$x \equiv x_1$$
$$\dot{x} \equiv x_2$$

Equation (3-21) is then written as

$$\dot{x}_1 = x_2 \qquad = f_1(\mathbf{x})$$
$$\dot{x}_2 = g(x_1, x_2) = f_2(\mathbf{x})$$

which is of the form

$$\dot{\mathbf{x}} = \mathbf{f}(\mathbf{x})$$

3.6 Concluding Remarks

The result established in this chapter allowing stability to small perturbations to be studied by linearization is the basis of a large number of studies and is the subject of the next chapter. In the subsequent chapter we shall return to the notion of stability regions for nonlinear systems to determine what further results can be obtained without the restriction of infinitesimal perturbations.

BIBLIOGRAPHICAL NOTES

The contents of the entire chapter, as well as much that follows, originate in the original 1892 Russian monograph by Liapunov, now available in English translation,

Liapunov, A. M., *Stability of Motion*, Academic Press, New York, 1966.

In older texts a variety of transliterations of the Russian spelling is used, including Liapounoff. Some of the results were obtained independently at the same time by Poincaré, and these two mathematicians may be considered the founders of modern nonlinear mechanics. The basic approach used here follows

Denn, M. M., *AIChE J.*, *16*, 670 (1970).

In that paper **B** is taken as the identity matrix, and the results regarding linearization are not sufficiently general. There are several good proofs of the fundamental linearization theorem in

Bellman, R., *Stability Theory of Differential Equations*, McGraw-Hill, New York, 1953.

The use of linearization is established in most textbooks on stability and nonlinear ordinary differential equations. See, for example,

Davis, H. T., *Introduction to Nonlinear Differential and Integral Equations*, Dover, New York, 1962.

Hahn, W., *Theory and Application of Liapunov's Direct Method*, Prentice-Hall, Englewood Cliffs, N.J., 1963.

Perlmutter, D. D., *Stability of Chemical Reactors*, Prentice-Hall, Englewood Cliffs, N.J., 1972.

Various definitions of stability are discussed by Hahn. See also an extremely well-written introductory book,

LaSalle, J., and S. Lefschetz, *Stability by Liapunov's Direct Method*, Academic Press, New York, 1961.

Stability
to Infinitesimal
Perturbations

4

4.1 Introduction

In this chapter we shall apply the results concerning stability to infinitesimal disturbances to some concrete examples, including the continuous flow stirred tank reactor and the stress distribution in an anisotropic fluid. We saw in Chapter 2 that both of these systems have multiple steady-state solutions.

There is a result from elementary algebra which we shall find useful. Consider the equation

$$\lambda^2 + a_1\lambda + a_2 = 0 \tag{4-1}$$

This can be factored to

$$[\lambda - \lambda_1][\lambda - \lambda_2] = \lambda^2 - [\lambda_1 + \lambda_2]\lambda + \lambda_1\lambda_2 = 0$$

where λ_1 and λ_2 are the roots. Then

$$a_1 = -[\lambda_1 + \lambda_2] \qquad a_2 = \lambda_1\lambda_2$$

If λ_1 and λ_2 are real and negative, then a_1 and a_2 will both be positive. If λ_1 and λ_2 are complex conjugates, then a_2 must be positive. The sum $\lambda_1 + \lambda_2$ is twice the real part of the root, so that a_1 will be positive if the real part of the root is negative. The converse of these statements can also be established, leading to the conclusion that *for the roots of Eq.* (4-1) *to have negative real parts it is necessary and sufficient that a_1 and a_2 be positive.*

This result does not generalize to polynomials of higher order. For a polynomial of nth order to have only roots with negative real parts it is necessary for all coefficients to be positive but not sufficient. The necessary and sufficient conditions for negative roots are known as the *Routh-Hurwitz criterion*, expressed in one form as follows:

From the coefficients of the equation

$$a_0 \lambda^n + a_1 \lambda^{n-1} + \cdots + a_{n-1}\lambda + a_n = 0 \tag{4-2}$$

form the n determinants

$$\Delta_i = \begin{vmatrix} a_1 & a_3 & a_5 & \cdots & & 0 \\ a_0 & a_2 & a_4 & \cdots & & 0 \\ 0 & a_1 & a_3 & \cdots & & 0 \\ 0 & a_0 & a_2 & \cdots & & 0 \\ 0 & 0 & a_1 & \cdots & & 0 \\ 0 & 0 & a_0 & \cdots & & 0 \\ \vdots & \vdots & & & \ddots & \vdots \\ & & & & & a_i \end{vmatrix} \tag{4-3}$$

$i = 1, 2, \ldots, n$. In this definition $a_k \equiv 0$, $k > n$. *The n roots of Eq.* (4-2) *will have negative real parts if and only if $\Delta_i > 0$, $i = 1, 2, \ldots, n$.* For the special case $n = 2$ we recover the condition that all coefficients must be positive.

4.2 CFSTR

The differential equations for a continuous flow stirred tank reactor with an irreversible first-order reaction were given in Sec. 2.3 as

$$\dot{x} = 1 - x - \alpha x e^{-\gamma/y} \qquad\qquad = f_1(x, y) \tag{4-4}$$

$$\dot{y} = [1 + \delta]\{\phi - y + \alpha\beta x e^{-\gamma/y}\} = f_2(x, y) \tag{4-5}$$

Here we are using x and y in place of x_1 and x_2. The matrix $\mathbf{A}$, Eq. (3-11), has elements

$$a_{11} = \frac{\partial f_1}{\partial x} = -1 - \alpha e^{-\gamma/y_s}$$

$$a_{12} = \frac{\partial f_1}{\partial y} = -\frac{\alpha \gamma x_s}{y_s^2} e^{-\gamma/y_s} = -\frac{\alpha \gamma}{\beta y_s^2} F(y_s)$$

$$\text{(4-6)}$$

$$a_{21} = \frac{\partial f_2}{\partial x} = [1 + \delta]\alpha \beta e^{-\gamma/y_s}$$

$$a_{22} = \frac{\partial f_2}{\partial y} = -[1 + \delta] + \frac{[1 + \delta]\alpha \gamma}{y_s^2} F(y_s)$$

All partial derivatives are evaluated at $x = x_s$, $y = y_s$. The function $F(y)$ is defined by Eq. (2-22).

The eigenvalues of $\mathbf{A}$, denoted by λ, are obtained from the equation

$$\begin{vmatrix} a_{11} - \lambda & a_{12} \\ a_{21} & a_{22} - \lambda \end{vmatrix} = [a_{11} - \lambda][a_{22} - \lambda] - a_{12} a_{21} = 0$$

or

$$\lambda^2 - [a_{11} + a_{22}]\lambda + [a_{11}a_{22} - a_{12} a_{21}] = 0 \qquad \text{(4-7)}$$

According to the criterion developed in the preceding section, the roots λ_1 and λ_2 of Eq. (4-7) will have negative real parts if and only if the following two criteria are satisfied:

$$a_{11}a_{22} - a_{12} a_{21} = [1 + \delta][1 - \alpha F'(y_s)] > 0$$

$$a_{11} + a_{22} = -2 + \alpha F'(y_s) - \delta \left[1 - \frac{\alpha \gamma}{y_s^2} F(y_s) \right] < 0$$

$F'(y)$ is defined by Eq. (2-23). These inequalities can be written as

$$\alpha F'(y_s) < 1 \qquad \text{(4-8)}$$

$$\alpha F'(y_s) < 2 + \delta \left[1 - \frac{\alpha \gamma}{y_s^2} F(y_s) \right] \qquad \text{(4-9)}$$

Inequalities (4-8) and (4-9) are necessary and sufficient conditions for a steady state to be stable to infinitesimal perturbations. Note that for an adiabatic reactor $\delta = 0$ and that the second inequality will always be satisfied when the first one is.

In Sec. 2.5 we established the *slope condition*, which required that when inequality (4-8) is violated a steady state must be unstable. On that basis we concluded, in agreement with experiment, that the middle steady state for the reactor is unstable to infinitesimal perturbations and cannot be maintained in practice. Thus, the data of Vejtasa and Schmitz in Fig. 1.3 show only two steady states for the adiabatic reactor, though the steady-state equations predict a third steady state which lies between the two.

At the high- and low-temperature steady states, where the slope condition is satisfied, the second inequality must hold as well. The data in Fig. 1.4 are for a nonadiabatic reactor with a unique steady state. For this reactor the slope condition (4-8) is satisfied, but the heat transfer coefficient is such that inequality (4-9) is violated. Since there is only one steady state and it is unstable, the system continuously cycles, even though the feed conditions are constant. The linearized (infinitesimal perturbation) analysis can tell us nothing quantitative about these oscillations, however.

4.3 Anisotropic Fluid

The equations for the orientation vector of an Ericksen anisotropic fluid in simple shear flow were shown in Sec. 2.6 to be

$$\dot{n}_1 = [\beta_1 + \beta_2 \, \Gamma n_1 n_2] n_1 + \tfrac{1}{2}[\beta_3 - 1]\Gamma n_2 = f_1(\mathbf{n}) \tag{4-10a}$$

$$\dot{n}_2 = [\beta_1 + \beta_2 \, \Gamma n_1 n_2] n_2 + \tfrac{1}{2}[\beta_3 + 1]n_1 \quad = f_2(\mathbf{n}) \tag{4-10b}$$

$$\dot{n}_3 = [\beta_1 + \beta_2 \, \Gamma n_1 n_2] n_3 \qquad\qquad = f_3(\mathbf{n}) \tag{4-10c}$$

We shall consider the stability of the steady state $\mathbf{n}_s = \mathbf{0}$, for which Newtonian behavior is obtained. The matrix $\mathbf{A} = \nabla \mathbf{f}(\mathbf{n}_s)$ is obtained by differentiating the right-hand sides of Eqs. (4-10) with respect to components of $\mathbf{n}$ and then setting $\mathbf{n} = \mathbf{0}$:

$$a_{11} = \frac{\partial f_1}{\partial n_1} = \beta_1 \qquad a_{12} = \frac{\partial f_1}{\partial n_2} = \frac{1}{2}(\beta_3 - 1)\Gamma \qquad a_{13} = \frac{\partial f_1}{\partial n_3} = 0$$

$$a_{21} = \frac{\partial f_2}{\partial n_1} = \frac{1}{2}(\beta_3 + 1)\Gamma \qquad a_{22} = \frac{\partial f_2}{\partial n_2} = \beta_1 \qquad a_{23} = \frac{\partial f_2}{\partial n_3} = 0$$

$$a_{31} = \frac{\partial f_3}{\partial n_1} = 0 \qquad a_{32} = \frac{\partial f_3}{\partial n_2} = 0 \qquad a_{33} = \frac{\partial f_3}{\partial n_3} = \beta_1$$

The eigenvalues λ_1, λ_2, λ_3 for this third-order system are obtained from the roots of the equation

$$\begin{vmatrix} \beta_1 - \lambda & \frac{1}{2}[\beta_3 - 1]\Gamma & 0 \\ \frac{1}{2}[\beta_3 + 1]\Gamma & \beta_1 - \lambda & 0 \\ 0 & 0 & \beta_1 - \lambda \end{vmatrix} = (\beta_1 - \lambda)[\lambda^2 - 2\lambda\beta_1 + \beta_1^2 - \tfrac{1}{4}(\beta_3^2 - 1)\Gamma^2]$$

$$= 0 \tag{4-11}$$

The eigenvalues are therefore $\lambda = \beta_1$ and the roots of the equation

$$\lambda^2 - 2\lambda\beta_1 + \beta_1^2 - \tfrac{1}{4}[\beta_3^2 - 1]\Gamma^2 = 0 \tag{4-12}$$

According to the criterion established in Sec. 4.1, the eigenvalues will have negative real parts if and only if the following two conditions are satisfied:

$$-2\beta_1 > 0$$
$$\beta_1^2 - \tfrac{1}{4}[\beta_3^2 - 1]\Gamma^2 > 0$$

The first of these inequalities gives the same result as the eigenvalue $\lambda = \beta_1$, namely that β_1 *must be negative*. Otherwise the steady state $\mathbf{n} = \mathbf{0}$ will always be unstable, even when there is no motion ($\Gamma = 0$). This would be thermodynamically impossible. The second inequality defines a critical shear rate

$$\Gamma^2 < \Gamma_c^2 = \frac{4\beta_1^2}{\beta_3^2 - 1} \tag{4-13}$$

Below the critical shear rate the steady state $\mathbf{n} = \mathbf{0}$ is stable to small perturbations, and Newtonian behavior can occur. At higher shear rates the orientation vector cannot remain in the state $\mathbf{n} = \mathbf{0}$, and non-Newtonian behavior must be observed according to the theory. From Eq. (2-50) it can be shown that a real solution for the non-Newtonian steady state exists only above the critical shear rate, and an identical stability analysis establishes that this state is stable.

4.4 Feedback Control

Stability is an overriding consideration in the design of feedback control systems. Many of the systems of interest in chemical processing have a structure

$$\dot{\mathbf{x}} = \mathbf{f}(\mathbf{x}) + \mathbf{b}(\mathbf{x})u \tag{4-14}$$

u is a scalar control variable which regulates performance in the neighborhood of the steady state and vanishes for $\mathbf{x} = \mathbf{x}_s$. In the chemical reactor equations, for example, u could be a variation in the heat transfer coefficient U, Eq. (2-7), effected by changes in the coolant flow rate.

A linear multivariable regulator has the form

$$u = \mathbf{k} \cdot [\mathbf{x} - \mathbf{x}_s] \qquad (4\text{-}15)$$

and Eq. (4-14) becomes

$$\dot{\mathbf{x}} = \mathbf{f}(\mathbf{x}) + \mathbf{b}(\mathbf{x})\mathbf{k} \cdot [\mathbf{x} - \mathbf{x}_s] \qquad (4\text{-}16)$$

The matrix $\mathbf{bk}$ has elements $b_i k_j$. For small perturbations Eq. (4-16) can be written as

$$\dot{\boldsymbol{\xi}} = [\mathbf{A} + \mathbf{bk}] \cdot \boldsymbol{\xi} + \mathbf{o}(\boldsymbol{\xi}) \qquad (4\text{-}17)$$

where $\mathbf{A} = \nabla \mathbf{f}(\mathbf{x}_s)$ and $\mathbf{b}$ is used to denote $\mathbf{b}(\mathbf{x}_s)$. Stability is therefore governed by the eigenvalues of $\mathbf{A} + \mathbf{bk}$. The design problem is to choose $\mathbf{k}$ for effective control within stability bounds.

The problem can be put in perspective by a concrete example. Consider the system

$$\dddot{x} = g(x, \dot{x}, \ddot{x}) + bu$$

or, equivalently,

$$\begin{aligned}
\dot{x}_1 &= x_2 \\
\dot{x}_2 &= x_3 \\
\dot{x}_3 &= g(\mathbf{x}) + bu
\end{aligned} \qquad (4\text{-}18)$$

u is to be proportional to the offset in x,

$$u = kx = kx_1 \qquad (4\text{-}19)$$

The matrix $\mathbf{A}$ then has the form

$$\mathbf{A} = \begin{pmatrix} 0 & 1 & 0 \\ 0 & 0 & 1 \\ a_{31} & a_{32} & a_{33} \end{pmatrix}$$

and the eigenvalue equation for $\mathbf{A} + \mathbf{b}k$ is

$$\begin{vmatrix} -\lambda & 1 & 0 \\ 0 & -\lambda & 1 \\ a_{31} + bk & a_{32} & a_{33} - \lambda \end{vmatrix} = -\lambda^3 + a_{33}\lambda^2 + a_{32}\lambda + [a_{31} + bk] = 0 \qquad (4\text{-}20)$$

According to the Routh-Hurwitz criterion, the eigenvalues will have negative real parts if and only if the following four conditions are satisfied:

$$a_{33} < 0 \qquad\qquad (4\text{-}20\text{a})$$

$$a_{32} < 0 \qquad\qquad (4\text{-}20\text{b})$$

$$a_{31} + bk < 0 \qquad\qquad (4\text{-}20\text{c})$$

$$a_{33}a_{32} + a_{31} > -bk \qquad\qquad (4\text{-}20\text{d})$$

From Eq. (4-20c) we see that stabilization is enhanced by having the product bk negative. This is the concept of negative feedback, and a system which is unstable because of a positive a_{31} can be stabilized with this control system. However, according to Eq. (4-20d), if $-bk$ is too large (too large a feedback gain, in control terminology), an instability will result which is *caused* by the control system. Thus, the control system plays the curious double role of stabilizing the system in one way and destabilizing it in another. The design range on the overall feedback gain, $-bk$, is therefore bounded from above and below.

4.5 Marginal Stability

When a parameter appears in the equations describing a dynamical process the concept of *marginal stability* becomes important. The point of marginal stability is defined as that value of the parameter for which an equilibrium will become unstable. Equivalently, if the solutions of the eigenvalue equation are considered to be functions of the parameter, then marginal stability occurs when the real part of an eigenvalue changes from negative (stable) to positive (unstable). This is sometimes also called the point of *neutral stability*.

In principle, we can obtain the point of marginal stability directly by setting the real part of the eigenvalue to zero in the eigenvalue equation and solving for the parameter. Equation (4-12) for the anisotropic fluid can be readily shown to have only real roots. Setting λ to zero we obtain the equation for the critical shear rate, Γ_c,

$$\beta_1^2 - \tfrac{1}{4}[\beta_3^2 - 1]\Gamma_c^2 = 0$$

which is consistent with the result obtained in Sec. 4.3.

Usually, it is necessary to consider the possibility of complex eigenvalues. For the control example in Sec. 4.5 the point of marginal stability for the feedback gain,

$$-bk_c = a_{33}a_{32} + a_{31}$$

is obtained by setting the real part of λ to zero in Eq. (4-20) but requiring that the imaginary part be nonzero. For the chemical reactor the important design parameter is α, the size parameter. It follows from inspection of Eqs. (4-4) and (4-5) or (4-6) that the system is stable as $\alpha \to 0$. Thus, there will be an upper bound on α for which stability occurs. By setting the real part of λ to zero in Eq. (4-7) we obtain two different values for α_c, depending on whether or not the imaginary part, λ_I, is zero:

$$\lambda_I = 0: \quad \alpha_c = \frac{1}{F'(y_s)}$$

$$\lambda_I \neq 0: \quad \alpha_c = \frac{2 + \delta[1 - (\alpha\gamma/y_s^2)F(y_s)]}{F'(y_s)}$$

When $\lambda_I \neq 0$ an instability is oscillatory, for the transient solution to the linearized equation contains terms $\sin \lambda_I t$, $\cos \lambda_I t$. A sufficient condition for real eigenvalues is that the matrix $\mathbf{A}$ be symmetric, but, as demonstrated by the anisotropic fluid equations, this condition is not necessary. When the eigenvalue equation has real roots the system is sometimes said to satisfy the *principle of exchange of stabilities*. Oscillatory instabilities are sometimes called *overstable*, since the unstable transient seems to arise from a tendency to return to the equilibrium point with such vigor that the system overshoots and continues on past with an even greater amplitude. The instability arising from too large a gain in a feedback control system typifies the overstable situation.

4.6 Concluding Remarks

Most stability analyses which have been carried out in the published literature are like those in this chapter, in which the stability of a steady state to infinitesimal perturbations is examined by use of the linearized dynamical equations. There are two important limitations to keep in mind:

1. No information is obtained about how large a perturbation can be tolerated before instability will occur. Thus, a steady state which is stable to infinitesimal perturbations might still be difficult to maintain in practice because it will be unstable to a small but finite perturbation.

2. Only information about asymptotic stability is obtained. A system with an asymptotically unstable steady state might still be stable with respect to regions that are quite small. In that case, the absence of a true steady state can be ignored in practice.

In the following two chapters we shall show how these limitations can be overcome to some extent. At present, however, the linear stability analysis remains the most useful single approach to system stability.

BIBLIOGRAPHICAL NOTES

A proof of the Routh-Hurwitz condition, with extensions and equivalent formulations, is given in either of the English translations of volume 2 of Gantmacher's excellent book,

Gantmacher, F. R., *Applications of the Theory of Matrices*, Wiley-Interscience, New York, 1959.

Gantmacher, F. R., *The Theory of Matrices*, Chelsea, New York, 1959.

Applications is a translation of volume 2 only. The original proof is in

Routh, E. J., *The Advanced Part of a Treatise on the Dynamics of a System of Rigid Bodies*, Dover, New York, 1955.

The linearized analysis of the continuous flow reactor was first carried out in

Bilous, O., and N. R. Amundson, *AIChE J.*, *1*, 513 (1955).

The extension to the stability of a controlled reactor is in

Aris, R., and N. R. Amundson, *Chem. Eng. Sci.*, *7*, 121, 132, 148 (1958).

More recent studies, including further experimental results, are tabulated in

Denn, M. M., *Ind. Eng. Chem.*, *61*, no. 2, 46 (1969).

Denn, M. M., in V. W. Weekman, Jr., ed., *Annual Review of Industrial and Engineering Chemistry*, *1970*, American Chemical Society, Washington, D.C., 1972.

Instabilities in biological reactors, such as those used in sewage treatment, are discussed in

Andrews, J. F., *Biotechnology and Bioengineering*, *10*, 707 (1968).

Yano, T., and S. Koga, *Biotechnology and Bioengineering*, *14*, 253 (1972).

Howell, J. A., C. T. Chi, and U. Pawlowsky, *Biotechnology and Bioengineering*, *14*, 253 (1972); *15*, 889, 897, 904 (1973).

For related stability problems in combustion, see

Frank-Kamenetskii, D. A., *Diffusion and Heat Transfer in Chemical Kinetics*, 2nd ed., Plenum, New York, 1969.

Chemical reaction on the surface of a catalytic wire also has a similar mathematical structure in the simplest approximation. See

Cardoso, M. A. A., and D. Luss, *Chem. Eng. Sci.*, *24*, 1699 (1969).

An instability may be the cause of the flickering which destroys catalytic wire gauze in systems like the commercial ammonia reactor. For a broad coverage of the area of chemical reaction stability, see

Perlmutter, D. D., *Stability of Chemical Reactors*, Prentice-Hall, Englewood Cliffs, N.J., 1972.

The stability analysis for transition to non-Newtonian behavior in the anisotropic fluid was given in

Ericksen, J. L., *Trans. Soc. Rheol.*, *6*, 275 (1962).

Stability of feedback control systems is treated in great detail in texts on process control, where practical methods for the design of stable control systems are developed. See, for example,

Coughanowr, D. R., and L. B. Koppel, *Process Systems Analysis and Control*, McGraw-Hill, New York, 1965.

Douglas, J. M., *Process Dynamics and Control*, Prentice-Hall, Englewood Cliffs, N.J., 1972.

Perlmutter, D. D., *Introduction to Chemical Process Control*, Wiley, New York, 1965.

Linearized stability analyses have been applied to many other systems of engineering interest. The stability of a fluid catalytic cracking unit is studied in

Iscol, L., *Preprints 1970 Joint Automatic Control Conference*, *Atlanta*, p. 602, Am. Soc. Mech. Eng., New York, 1970.

Lee, W., and A. M. Kugelman, *Ind. Eng. Chem. Process Design Develop.*, *12*, 197 (1973).

Iscol's paper relates the slow drifts seen in commercial cracking units to an instability. For a study of the stability of a continuous crystallizer, see

Sherwin, M. B., R. Shinnar, and S. Katz, *AIChE J.*, *6*, 114 (1967); *Chem. Eng. Progr. Symp. Ser. No. 95*, *65*, 75 (1969).

The very extensive literature on the stability of nuclear reactors is summarized in some recent texts,

Akcasu, Z., G. S. Lellouche, and L. M. Shotkin, *Mathematical Methods in Nuclear Reactor Dynamics*, Academic Press, New York, 1971.

Ash, M., *Nuclear Reactor Kinetics*, McGraw-Hill, New York, 1965.

Hetrick, D. L., *Dynamics of Nuclear Reactors*, University of Chicago Press, Chicago, 1971.

Several other interesting examples are given by Friedly,

Friedly, J. C., *Dynamic Behavior of Processes*, Prentice-Hall, Englewood Cliffs, N.J., 1972.

Liapunov's Direct Method

5

5.1 Introduction

We can obtain some insight into determining the size of the perturbation for which a steady state will remain stable by returning to the basic inequality (Sec. 3.2). If we can construct a family of surfaces $\mathscr{S}(\mathbf{x}, c) = 0$ which enclose the steady state such that $\mathbf{n} \cdot \mathbf{f} \le 0$, then we can ensure stability at least with respect to that family of surfaces. This is a sufficient condition for stability only, however. As we have noted previously, there might be perturbations whose transients leave the particular region and yet always return to the steady state.

We shall change nomenclature slightly here. It is traditional to define the family of surfaces by a positive definite function* $V(\xi)$

$$V(\xi) = \text{constant} \tag{5-1}$$

Here, ξ represents the deviation from steady state,

$$\xi = \mathbf{x} - \mathbf{x}_s \tag{5-2}$$

---

* A function is positive definite if

$$V(\xi) > 0, \quad \xi \ne 0$$
$$V(0) = 0$$

and the transient satisfies Eq. (3-9),

$$\dot{\xi} = \mathbf{f}(\mathbf{x}_s + \xi) \tag{5-3}$$

The normal to the family of surfaces is proportional to ∇V, so the basic inequality for stability is

$$\nabla V \cdot \mathbf{f} \leq 0 \qquad \text{everywhere on } V(\xi) = \text{constant} \tag{5-4}$$

Asymptotic stability requires strict inequality.

There is an alternative point of view, leading to the same result, which is helpful to consider. Constant values of the positive definite function $V(\xi)$ represent, in some sense, distance from the origin $\xi = 0$. By examining the value of $V(\xi)$ as time progresses we can determine whether or not $\xi(t)$ is moving closer to the origin. Thus,

$$\dot{V}(\xi(t)) = \sum_i \frac{\partial V}{\partial \xi_i} \dot{\xi}_i = \sum_i \frac{\partial V}{\partial \xi_i} f_i = \nabla V \cdot \mathbf{f} \tag{5-5}$$

If $\dot{V}$ is negative, then as time progresses the system moves to smaller and smaller values of V and hence closer to the origin. Thus, we obtain inequality (5-4) as a sufficient condition for stability. We may summarize this observation as *a sufficient condition for stability is that there exist a positive definite function $V(\xi)$ with a negative semidefinite (≤ 0) time derivative $\dot{V}(\xi)$. A sufficient condition for asymptotic stability is that $\dot{V}(\xi)$ be negative definite (<0, $\xi \neq 0$).*

A function which satisfies these definiteness conditions is called a *Liapunov function*, and this approach to stability is often called *Liapunov's direct* (or *second*) *method*. Note that if a Liapunov function can be found, we can establish stability within a region without linearization or the necessity of solving the nonlinear system differential equations. Furthermore, by reversing the arguments which have taken us to this point we can also establish the converse relation, *a sufficient condition for instability is that there exist a positive definite function $V(\xi)$ with a positive time derivative $\dot{V}(\xi)$*. Clearly, Liapunov's direct method is potentially very powerful. The problem in application is that it is often quite difficult to obtain a positive definite function $V(\xi)$ with a sign-definite (positive or negative) derivative. Failure to obtain a Liapunov function proves nothing about stability, but only that the family of surfaces which we have chosen is not a useful one.

5.2 Mechanical Energy

The use of a Liapunov function to establish stability is closely related to demonstrating that an equivalent mechanical process is dissipative. Consider the mass-spring-dashpot system in Fig. 5.1. ξ represents the displacement of

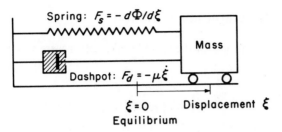

Figure 5.1 Schematic diagram of a mass-spring-dashpot system.

the mass from equilibrium. The spring is nonlinear, with a restoring force per unit mass

$$F_s = -\frac{d\Phi(\xi)}{d\xi} \tag{5-6}$$

$\Phi(\xi)$ is the potential energy per unit mass, a positive definite function. The viscous damping is assumed to be position- and velocity-dependent, so the drag force per unit mass is

$$F_d = -\mu(\xi, \dot{\xi})\dot{\xi} \tag{5-7}$$

The viscosity, $\mu(\xi, \dot{\xi})$, is positive. The dynamics are then described by the equation

$$\ddot{\xi} + \mu(\xi, \dot{\xi})\dot{\xi} + \frac{d\Phi(\xi)}{d\xi} = 0 \tag{5-8}$$

with a steady state at $\xi = \dot{\xi} = 0$.
    The total energy is the sum of kinetic and potential energy,

$$V = \tfrac{1}{2}\dot{\xi}^2 + \Phi(\xi) > 0 \tag{5-9}$$

The rate of change of energy is then

$$\dot{V} = \dot{\xi}\ddot{\xi} + \frac{d\Phi(\xi)}{d\xi}\dot{\xi} = -\mu(\xi, \dot{\xi})\dot{\xi}^2 < 0 \tag{5-10}$$

Thus, as long as the viscosity is positive the energy decreases with time and the system returns to equilibrium. The Liapunov function here is simply the total energy of the system.

5.3 Quadratic Liapunov Function

The dynamical equations for the stirred tank reactor were given in Sec. 2.3 as

$$\dot{x} = 1 - x - \alpha x e^{-\gamma/y} \qquad (5\text{-}11)$$

$$\dot{y} = [1 + \delta]\{\phi - y + \alpha\beta x e^{-\gamma/y}\} \qquad (5\text{-}12)$$

with the steady state x_s, y_s obtained by setting the time derivatives to zero. If we define deviation variables

$$\xi = x - x_s \qquad \eta = y - y_s$$

then the steady state corresponds to $\xi = \eta = 0$, and the dynamical equations are in the form of Eq. (5-3),

$$\dot{\xi} = -\xi - \alpha x_s\, \Delta(y, y_s)\eta - \alpha e^{-\gamma/y}\xi \qquad (5\text{-}13)$$

$$\dot{\eta} = [1 + \delta]\{-\eta + \alpha\beta x_s\, \Delta(y, y_s)\eta + \alpha\beta e^{-\gamma/y}\xi\} \qquad (5\text{-}14)$$

$$\Delta(y, y_s) \equiv \frac{e^{-\gamma/y} - e^{-\gamma/y_s}}{y - y_s} \qquad (5\text{-}15)$$

We have mixed the nomenclature here for shorthand purposes, often using y in place of $y_s + \eta$. Note that the function $\Delta(y, y_s)$ is positive and that

$$\lim_{y \to y_s} \Delta(y, y_s) = \frac{\gamma}{y_s^2} e^{-\gamma/y_s} \qquad (5\text{-}16)$$

The simplest family of positive definite functions is a sum of squares,

$$V = \tfrac{1}{2}[\xi^2 + \eta^2] \qquad (5\text{-}17)$$

Then

$$\begin{aligned}
\dot{V} &= \xi\dot{\xi} + \eta\dot{\eta} \\
&= \xi[-\xi - \alpha x_s\, \Delta(y, y_s)\eta - \alpha e^{-\gamma/y}\xi] \\
&\quad + \eta[1 + \delta][-\eta + \alpha\beta x_s\, \Delta(y, y_s)\eta + \alpha\beta e^{-\gamma/y}\xi] \\
&= -[1 + \alpha e^{-\gamma/y}]\xi^2 - \alpha\{x_s\, \Delta(y, y_s) - [1 + \delta]\beta e^{-\gamma/y}\}\xi\eta \\
&\quad - [1 + \delta][1 - \alpha\beta x_s\, \Delta(y, y_s)]\eta^2
\end{aligned} \qquad (5\text{-}18)$$

$\dot{V}$ is written as a quadratic form in ξ and η. It will be negative definite* if and only if the following two inequalities are satisfied:

$$1 + \alpha e^{-\gamma/y} > 0 \tag{5-19}$$

$$4[1 + \delta][1 + \alpha e^{-\gamma/y}][1 - \alpha\beta x_s \, \Delta(y, y_s)] - \alpha^2\{x_s \, \Delta(y, y_s) - [1 + \delta]\beta e^{-\gamma/y}\}^2 > 0 \tag{5-20}$$

The first of these inequalities is always satisfied, so the negative definiteness of $\dot{V}$ depends on the second, which is a function only of y. Let y_c be the first value of y for which inequality (5-20) is violated. The situation is then as shown in Fig. 5.2. Stability is ensured as long as ξ and η are initially within the

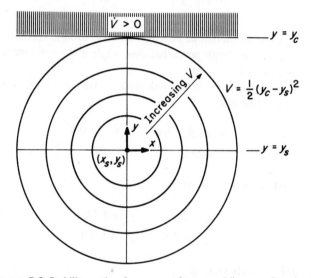

Figure 5.2 Stability region for a sum-of-squares Liapunov function.

family of circles which do not enter the shaded region, since only here is V positive definite and $\dot{V}$ negative definite. The limiting circle is the one tangent to the straight line $y = y_c$. According to Eq. (5-17), then, asymptotic stability is ensured for the region

$$[x - x_s]^2 + [y - y_s]^2 < [y_c - y_s]^2 \tag{5-21}$$

* A quadratic form $a_{11}\xi^2 + [a_{12} + a_{21}]\xi\eta + a_{22}\eta^2$ is positive definite if and only if

$$a_{11} > 0$$
$$4a_{11}a_{22} - [a_{12} + a_{21}]^2 > 0$$

and negative definite if the first inequality only is changed to $a_{11} < 0$.

Note that larger perturbations might still be stable, but we are unable to prove that with the particular choice of Liapunov function.

Inequality (5-20) can be quite conservative and might not even be satisfied at a stable steady state. In the limit $y \to y_s$, using Eq. (5-16), we obtain for an adiabatic reactor ($\delta = 0$)

$$\delta = 0, \qquad \lim_{y \to y_s}: \quad 1 - \frac{\alpha \beta \gamma x_s e^{-\gamma/y_s}}{y_s^2} > \frac{\alpha^2 e^{-2\gamma/y_s}[(x_s \gamma/y_s^2) - \beta]^2}{4[1 + \alpha e^{-\gamma/y_s}]} > 0$$

On the other hand, the slope condition, inequality (4-8), can be written as

$$1 - \frac{\alpha \beta \gamma x_s}{y_s^2} e^{-\gamma/y_s} > -\alpha e^{-\gamma/y_s}$$

Thus, the slope condition can be satisfied for an adiabatic reactor, establishing asymptotic stability to small perturbations, while a sum-of-squares Liapunov function may not exist.

Figure 5.3 shows the results of calculations for an adiabatic reactor with the following parameters:

$$\alpha = 1.8 \times 10^{10} \qquad \beta = 0.554$$
$$\gamma = 29.9 \qquad \delta = 0$$

There are stable steady states at

$$x_s = 0.014 \qquad y_s = 1.546$$
$$x_s = 0.998 \qquad y_s = 1.001$$

and an unstable steady state at

$$x_s = 0.532 \qquad y_s = 1.259$$

The dashed line in the figure separates the regions where each state is stable. The line is known as the *separatrix* and is obtained numerically from the full nonlinear equations. The circle with radius $|y_c - y_s| = 0.197$ about the low-temperature steady state is the stability region computed by searching for the largest region satisfying inequality (5-20). For these parameters the inequality is not satisfied anywhere about the stable high-temperature steady state. A finite stability region could be obtained about the high-temperature state and better results obtained about the low-temperature state by using a more general quadratic form,

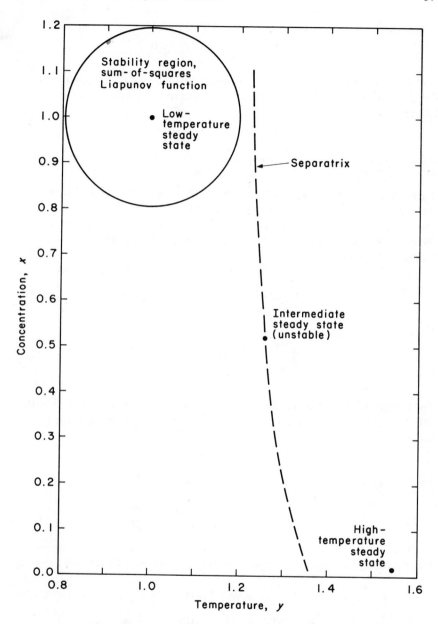

Figure 5.3 Regions of stability of high- and low-temperature steady states for an adiabatic continuous flow well-stirred reactor, with the stability region about the low-temperature steady state from a sum-of-squares Liapunov function.

$$V = \alpha_{11}\xi^2 + 2\alpha_{12}\,\xi\eta + \alpha_{22}\,\eta^2$$
$$\alpha_{11} > 0, \qquad \alpha_{11}\alpha_{22} - \alpha_{12}^2 > 0$$

and searching for the values of the parameters which give the largest region of stability. In general, however, quadratic Liapunov functions will be quite conservative.

5.4 Other Liapunov Functions

There is a very extensive literature on methods of constructing Liapunov functions for systems of various types. To give some feeling for the concepts employed we shall briefly discuss two of these methods. The first is known as the *method of Krasovskii*. For the system described by Eq. (5-3) we take as a Liapunov function

$$V(\xi) = \frac{1}{2}\mathbf{f}\cdot\mathbf{f} = \frac{1}{2}\sum_i f_i^2 > 0 \qquad (5\text{-}22)$$

Then

$$\dot{V}(\xi) = \sum_{i,j} f_i \frac{\partial f_i}{\partial \xi_j} f_j \qquad (5\text{-}23)$$

The derivative $\dot{V}$ is a quadratic form in $\mathbf{f}$ with coefficients $\partial f_i/\partial x_j$. For the stirred tank reactor equations (5-13) and (5-14) the quadratic form (5-23) becomes

$$\dot{V}(\xi) = -[1 + \alpha e^{-\gamma/y}]f_1^2 - \left\{\frac{\alpha\gamma x}{y^2} - [1 + \delta]\alpha\beta\right\}e^{-\gamma/y}f_1 f_2$$
$$- [1 + \delta]\left[1 - \frac{\alpha\beta\gamma x}{y^2}e^{-\gamma/y}\right]f_2^2 \qquad (5\text{-}24)$$

The necessary and sufficient conditions for negative definiteness are

$$1 + \alpha e^{-\gamma/y} > 0 \qquad (5\text{-}25)$$

$$4[1 + \delta][1 + \alpha e^{-\gamma/y}]\left[1 - \frac{\alpha\beta\gamma x}{y^2}e^{-\gamma/y}\right] - \alpha^2\left\{\frac{\gamma x}{y^2}e^{-\gamma/y} - [1 + \delta]\beta e^{-\gamma/y}\right\}^2 > 0$$
$$(5\text{-}26)$$

Inequality (5-25) is always satisfied.

Inequality (5-26) depends on x and y. As we approach the steady state it is readily shown that inequalities (5-20) and (5-26) approach the same limit, so when the quadratic Liapunov function gives a small stability region for the reactor no further useful information can be obtained from inequality (5-26). Thus, for the reactor parameters in the preceding section inequality (5-26) is not satisfied in any region about the high-temperature steady state. The largest stability region about the low-temperature steady state is found by maximizing V while keeping $\dot{V} < 0$ everywhere on the boundary $V = $ constant. This maximum occurs at a value $V = 0.00485$, giving the region shown in Fig. 5.4. Note that the quadratic form in $\mathbf{f}$ leads to a noncircular region in the ξ-η plane, since contours of Eq. (5-22) are not circles. Krasovskii's method can be generalized to functions of the form $V = \mathbf{f} \cdot \mathbf{Q} \cdot \mathbf{f}$, where $\mathbf{Q}$ is positive definite.

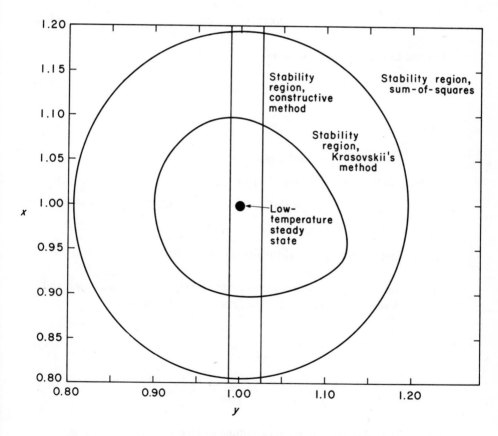

Figure 5.4 Stability regions about the low-temperature steady state computed by a sum-of-squares Liapunov function, Krasovskii's method, and the integral constructive method.

The second method which we shall examine is a constructive one involving unknown functions for the particular structure of the reactor equations. For algebraic simplicity we shall consider only the adiabatic reactor, $\delta = 0$. Two arbitrary functions are introduced, $r(\eta)$ and $q(\eta)$, and two integrals,

$$R(\eta) = \int_0^\eta r(\sigma)\, d\sigma \qquad\qquad (5\text{-}27a)$$

$$Q(\eta) = \int_0^\eta q(\sigma)\, d\sigma \qquad\qquad (5\text{-}27b)$$

$Q(\eta)$ is required to be positive definite, for which it is sufficient that $q(\eta)$ satisfy the inequality

$$\eta q(\eta) > 0, \qquad \eta \neq 0 \qquad\qquad (5\text{-}28)$$

For the Liapunov function we take

$$V = \tfrac{1}{2}[\xi - R(\eta)]^2 + Q(\eta) \qquad\qquad (5\text{-}29)$$

Then

$$\begin{aligned}
\dot{V} &= [\xi - R(\eta)][\dot{\xi} - r(\eta)\dot{\eta}] + q(\eta)\dot{\eta} \\
&= -\xi^2[1 + \alpha e^{-\gamma/y} + \alpha\beta e^{-\gamma/y} r(\eta)] \\
&\quad + \xi\{-\alpha x_s\, \Delta(y, y_s)\eta + [1 + \alpha e^{-\gamma/y}]R(\eta) + \alpha\beta e^{-\gamma/y}[q(\eta) + r(\eta)R(\eta)] \\
&\quad + [1 - \alpha\beta x_s\, \Delta(y, y_s)]r(\eta)\eta\} + \{\alpha x_s\, \Delta(y, y_s) \\
&\quad - [1 - \alpha\beta x_s\, \Delta(y, y_s)]r(\eta)\}R(\eta)\eta - [1 - \alpha\beta x_s\, \Delta(y, y_s)]q(\eta)\eta \qquad (5\text{-}30)
\end{aligned}$$

The derivative is made independent of ξ by setting the coefficients of ξ and ξ^2 to zero, leading to equations for the functions $r(\eta)$ and $q(\eta)$:

$$r(\eta) = -\frac{1 + \alpha e^{-\gamma/y}}{\alpha\beta e^{-\gamma/y}} \qquad\qquad (5\text{-}31)$$

$$q(\eta) = \frac{[1 + \alpha e^{-\gamma/y} - \alpha\beta x_s\, \Delta(y, y_s)]\eta}{[\alpha\beta e^{-\gamma/y}]^2} \qquad\qquad (5\text{-}32)$$

Inequality (5-28) and Eq. (5-32) combine to give

$$1 + \alpha e^{-\gamma/y} - \alpha\beta x_s\, \Delta(y, y_s) > 0 \qquad\qquad (5\text{-}33)$$

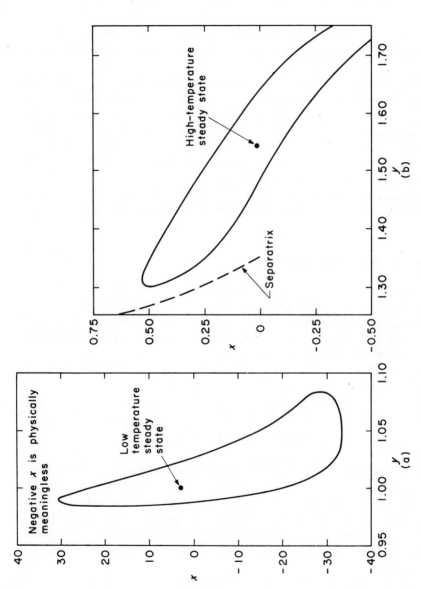

Figure 5.5 Stability regions about the (a) low-and (b) high-temperature steady states computed by the constructive integral method.

Equation (5-30) then simplifies to

$$\dot{V} = -\eta q(\eta)\left[1 + \alpha e^{-\gamma/y} - \alpha\beta x_s\,\Delta(y, y_s) + e^{-\gamma/y}\frac{1}{\eta}\int_0^\eta e^{\gamma/[y_s+\sigma]}\,d\sigma\right] \qquad (5\text{-}34)$$

Inequalities (5-28) and (5-33) are sufficient to ensure that $\dot{V}$ is negative definite. For the nonadiabatic reactor Eq. (5-34) provides an independent condition which must be satisfied.

The inequality (5-33) reduces to the slope condition as $y \to y_s$, and hence it will always give a finite stability region about a steady state which is asymptotically stable to small perturbations. The stability regions for the example problem, corresponding to $V = 244$ at the low-temperature steady state and $V = 0.00994$ at the high-temperature state, are shown in Fig. 5.5. The region about the low-temperature steady state is also shown on an expanded scale in Fig. 5.4. While the regions are quite large in total area in the temperature-concentration plane, the extent of the stability regions in the physically significant portion of the plane is disappointingly small.

5.5 Optimization Approach

The search for the *largest* region of stability for a given Liapunov function naturally suggests the use of optimization methods, and several such approaches have been developed. One which we shall also find useful subsequently in dealing with distributed parameter systems is based on the idea of marginal stability, first introduced in Sec. 4.5.

We shall suppose that there is a parameter, $\alpha > 0$, and that the system equations can be written as

$$\begin{aligned}
\dot{\xi} &= -\mathbf{C}\cdot\xi + \alpha\mathbf{B}(\xi)\cdot\xi \\
\dot{\xi}_i &= -\sum_j C_{ij}\xi_j + \alpha\sum_j B_{ij}(\xi)\xi_j
\end{aligned} \qquad (5\text{-}35)$$

We have called the parameter α so that in the subsequent example it will correspond to the size parameter in a flow reactor. The matrix $\mathbf{C}$ is positive definite, so the system is asymptotically stable as $\alpha \to 0$. For simplicity in the subsequent manipulations we shall take $\mathbf{C}$ as constant. This not necessary, but it is convenient and $\mathbf{C}$ is in fact constant in the chemical reactor equations. We pose the problem as seeking the largest value of α for which stability can be ensured.

We shall use a quadratic Liapunov function,

$$V = \frac{1}{2}\xi\cdot\xi = \frac{1}{2}\sum_i \xi_i^2$$

This is restrictive but convenient for illustration. The derivative is

$$\dot{V} = \xi \cdot \dot{\xi} = -\xi \cdot \mathbf{C} \cdot \xi + \alpha \xi \cdot \mathbf{B}(\xi) \cdot \xi \qquad (5\text{-}36)$$

We have asymptotic stability as long as $\dot{V}$ is negative, or, equivalently,

$$\alpha \xi \cdot \mathbf{B}(\xi) \cdot \xi < \xi \cdot \mathbf{C} \cdot \xi \qquad (5\text{-}37)$$

α and $\xi \cdot \mathbf{C} \cdot \xi$ are both positive, so we may divide by both without changing the sign of the inequality. Thus,

$$\frac{1}{\alpha} > \frac{\xi \cdot \mathbf{B}(\xi) \cdot \xi}{\xi \cdot \mathbf{C} \cdot \xi} \qquad (5\text{-}38)$$

The inequality is trivially satisfied if $\mathbf{B}(\xi)$ is negative definite, in which case the system is stable for all α.

For the system to be stable $1/\alpha$ must be larger than the right-hand side of inequality (5-38). Thus, if we define the maximum of the ratio $\xi \cdot \mathbf{B}(\xi) \cdot \xi / \xi \cdot \mathbf{C} \cdot \xi$ as $1/\lambda$, we can write

$$\frac{1}{\alpha} > \frac{1}{\lambda} = \max_{\xi} \frac{\xi \cdot \mathbf{B}(\xi) \cdot \xi}{\xi \cdot \mathbf{C} \cdot \xi} \qquad (5\text{-}39)$$

The maximum can be obtained by setting derivatives to zero. If we let $\mathbf{y}$ be the value of ξ at the maximum, we obtain the equation

$$\sum_{j} [C_{ij} + C_{ji}] y_j = \lambda \left\{ \sum_{j} [B_{ij} + B_{ji}] y_j + \sum_{j,k} y_j \frac{\partial B_{jk}}{\partial \xi_i} y_k \right\} \qquad (5\text{-}40)$$

Equation (5-40) is a nonlinear eigenvalue equation which always admits the trivial solution $\mathbf{y} = \mathbf{0}$. If there are no positive eigenvalues λ, the maximum in Eq. (5-34) is always negative, and global stability is ensured. If there is at least one positive eigenvalue, then the critical value of α is equal to the smallest positive eigenvalue.

In the limit $|\mathbf{y}| \to 0$, Eq. (5-40) reduces to a conventional linear eigenvalue problem,

$$\sum_{j} [C_{ij} + C_{ji}] y_j = \lambda \sum_{j} [B_{ij}(\mathbf{0}) + B_{ji}(\mathbf{0})] y_j$$

$$[\mathbf{C} + \mathbf{C}^T] \cdot \mathbf{y} = \lambda [\mathbf{B}(\mathbf{0}) + \mathbf{B}^T(\mathbf{0})] \cdot \mathbf{y} \qquad (5\text{-}41)$$

and the eigenvalues are found from the characteristic equation

$$|[\mathbf{C} + \mathbf{C}^T] - \lambda [\mathbf{B}(\mathbf{0}) + \mathbf{B}^T(\mathbf{0})]| = 0 \qquad (5\text{-}42)$$

Note that Eq. (5-42) is symmetric (self-adjoint) with all real eigenvalues. It is the symmetric part of the eigenvalue equation which results from linearizing Eq. (5-35). Call the smallest positive eigenvalue of this linear equation (5-42) λ_0 and let $\mathbf{y}_0$ be the corresponding eigenvector, normalized such that

$$|\mathbf{y}_0| = \left| \sum_i y_{0i}^2 \right|^{1/2} = 1 \qquad (5\text{-}43)$$

We can obtain a perturbation solution to the nonlinear equation (5-40) (Appendix B), which, to first order in $|\mathbf{y}_0|$, is

$$\mathbf{y} \simeq \mathscr{A}\mathbf{y}_0$$

$$\lambda \simeq \lambda_0 \left\{ 1 - \left[\frac{3 \sum\limits_{i,j,k} \gamma_{ijk}\, y_{0i}\, y_{0j}\, y_{0k}}{\sum\limits_{i,j} [B_{ij}(0) + B_{ji}(0)] y_{0i}\, y_{0j}} \right] \mathscr{A} \right\}$$

$$\gamma_{ijk} = \frac{\partial B_{jk}}{\partial \xi_i} \text{ at } \xi = 0$$

That is, for the nonlinear problem the eigenvalue λ depends on the norm of the eigenfunction. Inequality (5-39) then becomes

$$\alpha < \lambda_0 \left\{ 1 - \left[\frac{3 \sum\limits_{i,j,k} \gamma_{ijk}\, y_{0i}\, y_{0j}\, y_{0k}}{\sum\limits_{i,j} [B_{ij}(0) + B_{ji}(0)] y_{0i}\, y_{0j}} \right] \mathscr{A} \right\} \qquad (5\text{-}44)$$

The coefficient of $\mathscr{A}$ in Eq. (5-44) can be positive or negative, depending on the direction chosen for $\mathbf{y}_0$. A conservative bound on α is obtained by choosing the coefficient to be positive, or, more simply, always using the absolute value.

Inequality (5-44) can be rearranged to give a result which is more useful to us,

$$\mathscr{A} < \left| \frac{\sum\limits_{i,j} [B_{ij}(0) + B_{ji}(0)] y_{0i}\, y_{0j}}{\sum\limits_{i,j,k} \gamma_{ijk}\, y_{0i}\, y_{0j}\, y_{0k}} \right| \left[1 - \frac{\alpha}{\lambda_0} \right] \qquad (5\text{-}45)$$

The right-hand side of the inequality defines the limiting radius for which stability is ensured according to the quadratic Liapunov function $V = \frac{1}{2}\xi \cdot \xi$.

Thus, by this approach, we can obtain an estimate of the size of the stability region by solution of a linear eigenvalue problem, Eq. (5-42).

For the adiabatic stirred reactor we obtain from Eqs. (5-13) and (5-14)

$$\mathbf{C} = \begin{pmatrix} 1 & 0 \\ 0 & 1 \end{pmatrix}$$

$$\mathbf{B}(0) = e^{-\gamma/y_s} \begin{pmatrix} -1 & -\dfrac{\gamma x_s}{y_s^2} \\[2ex] \beta & \dfrac{\beta x_s \gamma}{y_s^2} \end{pmatrix}$$

$$\gamma_{1jk} = 0$$

$$\gamma_{2jk} = e^{-\gamma/y_s} \begin{pmatrix} -\dfrac{\gamma}{y_s^2} & -\dfrac{\gamma x_s}{2y_s^3}\left[\dfrac{\gamma}{y_s} - 2\right] \\[2ex] \beta \dfrac{\gamma}{y_s^2} & \dfrac{\beta x_s \gamma}{2y_s^3}\left[\dfrac{\gamma}{y_s} - 2\right] \end{pmatrix}$$

The linear estimate of marginal stability is obtained by solving Eq. (5-42),

first approximation: $\alpha < \lambda_0 = \dfrac{1}{F'(y_s)} \dfrac{2}{1 + \left\{1 + \dfrac{[\beta + \gamma x_s/y_s^2]^2}{[1 - \beta x_s \gamma/y_s^2]^2}\right\}^{1/2}}$ (5-46)

This result is always more conservative than application of the slope condition. For the parameters introduced in Sec. 5-3, Eqs. (5-45) and (5-46) give a value for the stability radius about the low-temperature steady state of

$$\mathscr{A} = \{[\xi - \xi_s]^2 + [y - y_s]^2\}^{1/2} < 0.162$$

This compares quite favorably with the value 0.197 obtained directly from the quadratic Liapunov function in Sec. 5-3. We have already found that the quadratic Liapunov function cannot be applied to the high-temperature steady state for the parameters in this example problem.

The primary importance of this approach lies in the fact that *an estimate of a finite stability region is obtained from solution of a linear eigenvalue problem.* For systems of higher order, and in particular distributed parameter systems, the linear problem can be solved without great difficulty, while a systematic study in the high- (or infinite-) dimensional space to determine the boundaries of the stability region would be difficult or impossible.

5.6 Feedback Control

Liapunov's direct method can be used in the design of multivariable feedback control systems. For simplicity we shall suppose that we are operating in a region near the steady state where linearized equations are sufficiently accurate. The linear equivalent of Eq. (4-14) is then

$$\dot{\xi} = \mathbf{A} \cdot \xi + \mathbf{b}u(\xi) \tag{5-47}$$

where $\mathbf{A}$ and $\mathbf{b}$ have constant elements and u is a scalar feedback control variable. We assume that the variables are normalized so that the possible range of control action is

$$|u| \le 1 \tag{5-48}$$

We shall also assume that the steady state is stable for $u = 0$, so that the sole function of the control system is to speed response.

Consider a Liapunov function

$$V = \tfrac{1}{2}\xi \cdot \xi$$

V is a measure of distance from the steady state. The derivative is

$$\dot{V} = \xi \cdot \mathbf{A} \cdot \xi + \mathbf{b} \cdot \xi u \tag{5-49}$$

where the first term is negative because of the assumption of a stable steady state. Any control which makes the term $\mathbf{b} \cdot \xi u$ negative enhances the stability and drives the system toward $V = 0$ more rapidly than the uncontrolled system. By minimizing $\dot{V}$ we shall obtain the control which causes the most rapid return. Because $\dot{V}$ is linear in u, it is made as negative as possible by the control

$$u = \begin{cases} +1, & \mathbf{b} \cdot \xi < 0 \\ -1, & \mathbf{b} \cdot \xi > 0 \end{cases}$$

or, equivalently,

$$u = -\frac{\mathbf{b} \cdot \xi}{|\mathbf{b} \cdot \xi|} \tag{5-50}$$

That is, the linear combination $\mathbf{b} \cdot \xi$ is a *switching function* whose algebraic sign determines which of its extreme values the control variable should take. $\dot{V}$ is then

$$\dot{V} = \xi \cdot \mathbf{A} \cdot \xi - |\mathbf{b} \cdot \xi| \tag{5-51}$$

For the stirred tank reactor, Eqs. (2-6) and (2-7), the heat transfer occurs only in the temperature equation ($b_1 = 0$), and the switching criterion is simply whether the temperature is above or below the steady-state temperature.

A control system of the type described by Eq. (5-50), which switches between its extreme values, is known as a *bang-bang*, or *relay*, controller. The control philosophy guarantees a stable, vigorous control system. It does have one weakness, however. By concentrating on driving V to zero as rapidly as possible we neglect the future consequences of the action, and more restrained action initially might allow for a better overall performance. In this particular case it is possible to show that the control (5-50) minimizes the *overall performance function*

$$\mathscr{E} = \int_0^\infty \{ -\boldsymbol{\xi} \cdot \mathbf{A} \cdot \boldsymbol{\xi} + |\mathbf{b} \cdot \boldsymbol{\xi}| \}\, dt \qquad (5\text{-}52)$$

and hence is a rigorously optimal control. A more general quadratic Liapunov function will lead to a more general quadratic performance function in Eq. (5-52).

5.7 Sufficient Condition for Instability

There is a method due to W. Eckhaus for estimating the size of a perturbation which causes instability which, though not strictly an application of Liapunov's direct method, is in the same spirit and is appropriately discussed here. One of the key assumptions will prove to be highly restrictive for lumped parameter systems, and little useful application is to be expected. The same assumption is a good one for distributed parameter systems, however; and the Eckhaus approach will be applied subsequently. It is helpful to introduce it here in the simpler context of lumped systems.

We shall suppose that the nonlinear equations have been expanded about the steady state through terms of *second order*:

$$\dot{\xi}_i = \sum_j A_{ij} \xi_j + \sum_{j,k} B_{ijk} \xi_j \xi_k + \cdots \qquad (5\text{-}53)$$

The steady state is stable to infinitesimal perturbations, so all eigenvalues of $\mathbf{A}$ have negative real parts. We shall assume for simplicity that all eigenvalues are real and distinct. It is helpful to recall some properties of the linear approximation to Eq. (5-53),

$$\dot{\xi}_i = \sum_j A_{ij} \xi_j \qquad (5\text{-}54)$$

which has a solution

$$\xi_i = \sum_n C_n e^{\lambda_n t} y_{ni} \tag{5-55}$$

Here λ_n is the eigenvalue, y_n is the corresponding eigenvector, and the C_n are constants determined by the initial conditions. The eigenvalue equation is

$$\sum_j A_{ij} y_{nj} = \lambda_n y_{ni} \tag{5-56}$$

We shall take each eigenvector as normalized to unity:

$$\mathbf{y}_n \cdot \mathbf{y}_n = \sum_i y_{ni} y_{ni} = 1 \tag{5-57}$$

Also important is the *adjoint* linear system,

$$\dot{\eta}_j = \sum_i \eta_i A_{ij} \tag{5-58}$$

This has a solution

$$\eta_j = \sum_n C_n^A e^{\lambda_n t} y_{nj}^A \tag{5-59}$$

$$\sum_i y_{ni}^A A_{ij} = \lambda_n y_j^A \tag{5-60}$$

The eigenvalues of Eqs. (5-56) and (5-60) are readily shown to be identical, while the eigenvectors are orthogonal:

$$\mathbf{y}_n \cdot \mathbf{y}_m^A = \sum_i y_{ni} y_{mi}^A = 0, \qquad n \neq m \tag{5-61a}$$

When $n = m$ the inner product is not zero, and the adjoint system can be normalized so that the product is unity:

$$\mathbf{y}_n \cdot \mathbf{y}_n^A = \sum_i y_{ni} y_{ni}^A = 1 \tag{5-61b}$$

The eigenvectors of Eq. (5-56) are linearly independent, so we can write any vector of the same dimension as a linear combination of them. We do so for the solution to Eq. (5-53),

$$\xi_i = \sum_n \mathscr{A}_n(t) y_{ni} \tag{5-62}$$

Substituting into Eq. (5-53) gives

$$\dot{\xi}_i = \sum_n \mathscr{A}_n y_{ni} = \sum_j \sum_n A_{ij} \mathscr{A}_n y_{nj} + \sum_{j,k} \sum_{n,m} B_{ijk} \mathscr{A}_n y_{nj} \mathscr{A}_m y_{mk} + \cdots \quad (5\text{-}63)$$

We now multiply by y_{pi}^A and sum on i, making use of Eqs. (5-61):

$$\dot{\mathscr{A}}_p = \sum_n \sum_j \sum_i y_{pi}^A A_{ij} y_{nj} \mathscr{A}_n + \sum_{n,m} I_{pnm} \mathscr{A}_n \mathscr{A}_m + \cdots \quad (5\text{-}64)$$

$$I_{pnm} \equiv \sum_{i,j,k} B_{ijk} y_{pi}^A y_{nj} y_{mk}$$

The first term on the right of Eq. (5-64) simplifies by use, sequentially, of Eqs. (5-60) and (5-61):

$$\sum_n \sum_j \left[\sum_i y_{pi}^A A_{ij} \right] y_{nj} \mathscr{A}_n = \sum_n \lambda_p \left[\sum_j y_{pj}^A y_{nj} \right] \mathscr{A}_n = \lambda_p \mathscr{A}_p$$

and Eq. (5-64) becomes, finally,

$$\dot{\mathscr{A}}_p = \lambda_p \mathscr{A}_p + \sum_{n,m} I_{pnm} \mathscr{A}_n \mathscr{A}_m + \cdots \quad (5\text{-}65)$$

Equation (5-65) is completely equivalent to the original Eq. (5-53), except that it is written in terms of the modes of the linear systems. At the linear approximation the system is uncoupled, and we simply obtain the result that $\mathscr{A}_p(t)$ behaves like $\exp(\lambda_p t)$. It is at this point that we introduce the Eckhaus assumptions to uncouple the higher-order terms. Note that the eigenvalues can be ordered

$$0 > \lambda_1 > \lambda_2 > \cdots$$

We shall write the terms in Eq. (5-65) to be of comparable order,

$$\mathscr{A}_p = \varepsilon \delta_p a_p \quad (5\text{-}66)$$

Here, ε represents the order of magnitude of the perturbation, δ_p the relative ordering of the coefficients, and a_p a term of order unity. We choose

$$\varepsilon = -\lambda_1 \qquad \delta_p = \frac{\lambda_1}{\lambda_p} \quad (5\text{-}67)$$

Then Eq. (5-65) can be written as

$$\dot{a}_p = \lambda_p a_p - \lambda_p \sum_{n,m} I_{pnm} \left(\frac{\lambda_1^2}{\lambda_n \lambda_m} \right) a_n a_m + \cdots \quad (5\text{-}68)$$

We shall seek a solution which is valid near the point of marginal stability, where $\lambda_1 \to 0$. In the summation in Eq. (5-68) the only term which will not vanish in that limit is $n = m = 1$, so that Eq. (5-68) becomes

$$\dot{a}_p = \lambda_p a_p - \lambda_p I_{p11} a_1^2 + \cdots$$

or, equivalently,

$$\dot{\mathscr{A}}_p = \lambda_p \mathscr{A}_p + I_{p11} \mathscr{A}_1^2 + \cdots \tag{5-69}$$

Note that the limiting process makes sense only if the eigenvalues are widely separated. This is the assumption which makes the procedure generally inapplicable to lumped systems, though it has been successfully applied to a lumped model of a batch catalytic fluidized bed.

Equation (5-69) is now essentially uncoupled, in that the dynamical behavior of all modes is forced by the behavior of the critical (first) mode. If $\mathscr{A}_1(t) \to 0$, then the forcing term vanishes and the linear lumped exponential behavior dominates. Conversely, if $\mathscr{A}_1(t)$ does not go to zero, then all $\mathscr{A}_p(t)$ will be forced away from zero. The stability of the system (5-53) then depends on the behavior of the equation for $\mathscr{A}_1(t)$,

$$\dot{\mathscr{A}}_1 = \lambda_1 \mathscr{A}_1 + I_{111} \mathscr{A}_1^2 + \cdots \tag{5-70}$$

This has a solution

$$\mathscr{A}_1(t) = \frac{\mathscr{A}_1(0)e^{\lambda_1 t}}{1 + \mathscr{A}_1(0)(I_{111}/\lambda_1)[1 - e^{\lambda_1 t}]} \tag{5-71}$$

where $\mathscr{A}_1(0)$ is the initial value of A_1. When the product $-\mathscr{A}_1(0)I_{111}/\lambda_1$ is less than unity, $\mathscr{A}_1(t)$ will decay with time approximately according to the linear decay rate. When the product is greater than unity, however, $\mathscr{A}_1(t)$ will grow without bound, and the system is unstable. There is, therefore, a critical amplitude,

$$\mathscr{A}_{1c}^2 = \left| \frac{\lambda_1}{I_{111}} \right|^2 \tag{5-72}$$

beyond which instability will occur. Because of the assumptions on the ordering of terms, $\mathscr{A}_{1c}^2$ is approximately equal to the critical amplitude $|\xi|^2$ for Eq. (5-53).

Within the accuracy of truncation at second order and spacing of the eigenvalues, Eq. (5-72) represents a sufficient condition for instability near the point of marginal stability, requiring only the solution of a linear eigenvalue problem and its adjoint. There is also an implicit assumption that the form of the initial disturbance is dominated by the first eigenvector, and some other disturbance might cause instability at a smaller amplitude. Thus, we have a situation very similar to the converse of Liapunov's direct method, where we have established a condition for a growing disturbance at some distance from a steady state which is stable to infinitesimal disturbances.

5.8 Concluding Remarks

Liapunov's direct method has attracted considerable attention in stability studies because of its elegance and potential power. The ability to analyze the finite-amplitude behavior of a nonlinear system without ever solving the system differential equations is extremely attractive. The main problem in application has been the general inability to construct Liapunov functions which define stability regions of a sufficient size to be useful in practice. Encouraging results are available for problems with special structure, but, on the whole, the practical return for the large research investment in the 1960's is disappointing. We are more optimistic about the utility of the method for analyzing the behavior of distributed parameter systems. This is dealt with in Chapter 15.

BIBLIOGRAPHICAL NOTES

The fundamental source is

Liapunov, A. M., *Stability of Motion*, Academic Press, New York, 1966.

The direct method is dealt with comprehensively in a number of other texts, including

Hahn, W., *Theory and Application of Liapunov's Direct Method*, Prentice-Hall, Englewood Cliffs, N.J., 1963.

Hahn, W., *Stability of Motion*, Springer-Verlag, Inc., New York, New York, 1967.

LaSalle, J., and S. Lefschetz, *Stability by Liapunov's Direct Method*, Academic Press, New York, 1961.

Zubov, V. I., *Methods of A. M. Lyapunov and Their Application*, Noordhoff, Groningen, Netherlands, 1964.

Useful survey papers are

Gurel, O., and L. Lapidus, *Ind. Eng. Chem.*, 60, no. 6, 13 (1968).

Szegö, G. P., *Appl. Mech. Rev.*, 19, 833 (1966).

The connection with a dissipative mechanical system is mentioned in most texts. For the relation of Liapunov's direct method to the second law of thermodynamics, see

> Coleman, B. D., and V. Mizel, *Arch. Rat. Mech. Anal.*, *25*, 243 (1967).
> Glansdorff, P., and I. Prigogine, *Thermodynamic Theory of Structure, Stability and Fluctuation*, Wiley, New York, 1972.
> Wei, J., *J. Chem. Phys.*, *36*, 1578 (1962).

For a general theory of dissipative dynamical systems in which the Liapunov function is related to storage and supply functions, see

> Willems, J. C., *Arch. Rat. Mech. Anal.*, *45*, 321, 352 (1973).

The application of Liapunov's direct method to chemical reactors is covered quite broadly in

> Perlmutter, D. D., *Stability of Chemical Reactors*, Prentice-Hall, Englewood Cliffs, N.J., 1972.

Application to nuclear reactors is given in the texts cited in Chapter 4, particularly,

> Hetrick, D. L., *Dynamics of Nuclear Reactors*, University of Chicago Press, Chicago, 1971.

The constructive method in Sec. 5-4 follows

> Warden, R. B., R. Aris, and N. R. Amundson, *Chem. Eng. Sci.*, *19*, 149 (1964).

Methods of construction of Liapunov functions are surveyed in

> Gurel, O., and L. Lapidus, *Ind. Eng. Chem.*, *61*, no. 3, 30 (1969).
> Schultz, D. G., in C. T. Leondes, ed., *Advances in Control Systems*, vol. 2, Academic Press, New York, 1965.

The optimization approach in Sec. 5-5 is based on a similar method developed for distributed parameter systems. References are given at the appropriate point in Chapter 15.

The use of Liapunov's direct method for the construction of optimal control systems, with references to earlier work, is covered in

> Denn, M. M., *Optimization by Variational Methods*, McGraw-Hill, New York, 1969.
> Lapidus, L., and R. Luus, *Optimal Control of Engineering Processes*, Ginn/Blaisdell, Waltham, Mass., 1967.

The stability of feedback control systems, including extensions of the Liapunov method by Popov, is treated in

Lefschetz, A., *Stability of Nonlinear Control Systems*, Academic Press, New York, 1965.

Minorsky, N., *Theory of Nonlinear Control Systems*, McGraw-Hill, New York, 1969.

The Eckhaus method was developed for distributed parameter systems and complete references are given in Chapter 17. The basic reference is

Eckhaus, W., *Studies in Non-Linear Stability Theory*, Springer-Verlag, New York, Inc., New York, 1965.

It has been applied successfully to a lumped parameter model of a batch catalytic fluidized bed in an unpublished paper by Fisher and Denn. The model of the fluidized bed is in

Luss, D., and N. R. Amundson, *AIChE J.*, *14*, 211(1968).

The decomposition into normal modes is closely related to the idea of modal control. See, for example,

Gould, L. A., *Chemical Process Control: Theory and Applications*, Addison-Wesley, Reading, Mass., 1969.

Nonlinear Oscillations

6

6.1 Introduction

Steady oscillations in an unforced system, such as those described in Sec. 1.2 in a chemical reactor, are a consequence of the nonlinear interactions in the system. A detailed quantitative description of such phenomena will always require computer simulation of the nonlinear model. Under certain limiting conditions, however, approximate solutions to the nonlinear differential equations can be obtained. Such approximate solutions are useful because they provide an analytical description of the effect of system parameters on process behavior and hence define the regions of interest for more detailed investigation.

The starting point in the study of nonlinear oscillations is with the linear oscillator,

$$\ddot{u} + \omega_0^2 u = 0 \tag{6-1}$$

with the solution

$$u(t) = \mathscr{A} \cos(\omega_0 t + \phi) \tag{6-2}$$

Because of the linearity the amplitude $\mathscr{A}$ is arbitrary and is uniquely specified, along with the phase angle ϕ, only when two initial conditions are given.

As in Sec. 5-2, we form the total energy by multiplying Eq. (6-1) by $\dot{u}$ and integrating,

$$\dot{u}\ddot{u} + \omega_0^2 \dot{u}u = \frac{1}{2}\frac{d}{dt}[\dot{u}^2 + \omega_0^2 u^2] = 0$$

$$\text{energy} = \frac{1}{2}[\dot{u}^2 + \omega_0^2 u^2] = \text{constant} \qquad (6\text{-}3)$$

The oscillations are a consequence of the conservative nature of the system. In the u-$\dot{u}$ plane all trajectories lie on a closed elliptical curve (Fig. 6.1). Thus, when the system returns to a given position, u, it returns also to the corresponding value of $\dot{u}$, and hence the periodic output. The amplitude of the ellipse is uniquely determined by the initial conditions.

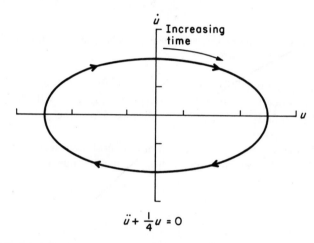

Figure 6.1 Trajectory for the system $\ddot{u} + \omega_0^2 u = 0$. $\omega_0 = 1/2$ for the curve shown here.

A nonlinear conservative system behaves in essentially the same way. If the potential energy is $\Phi(u)$, then the system equation is

$$\ddot{u} + \frac{d\Phi(u)}{du} = 0 \qquad (6\text{-}4)$$

Multiplying by $\dot{u}$ and integrating gives

$$\dot{u}\ddot{u} + \frac{d\Phi(u)}{du}\dot{u} = \frac{d}{dt}\left[\frac{1}{2}\dot{u}^2 + \Phi(u)\right] = 0$$

$$\text{energy} = \frac{1}{2}\dot{u}^2 + \Phi(u) = \text{constant} \qquad (6\text{-}5)$$

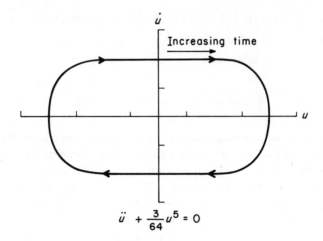

Figure 6.2 Trajectory for the system $\ddot{u} + d\Phi(u)/du = 0$. $\Phi(u) = u^6/128$ for the curve shown here.

As long as the potential $\Phi(u)$ is a positive definite function the system describes closed (nonelliptical) contours in the u-$\dot{u}$ plane (Fig. 6.2) with the amplitude uniquely determined by the initial conditions. Thus, the output is periodic but no longer sinusoidal.

A condition for oscillations in a more general second-order system can now be defined. For consistency with subsequent usage we shall write the equation in the form

$$\ddot{u} + \frac{d\Phi(u)}{du} = \varepsilon\psi(u, \dot{u}) \tag{6-6}$$

where $\Phi(u)$ is positive definite. Multiplying by $\dot{u}$ gives

$$\frac{d}{dt}\left[\tfrac{1}{2}\dot{u}^2 + \Phi(u)\right] = \varepsilon\dot{u}\psi(u, \dot{u}) \tag{6-7}$$

It is convenient here to integrate over one period. In that case the term on the left vanishes because of the periodicity, and we obtain the condition

$$\varepsilon\int_0^{2\pi/\omega} \dot{u}(t)\psi(u, \dot{u})\, dt = 0 \tag{6-8}$$

where we have taken the frequency as ω and the corresponding period of oscillations as $2\pi/\omega$. The analytical problem, then, is to find conditions under which solutions to Eq. (6-6) satisfy Eq. (6-8). The remarkable phenomenon which results from the presence of the $\dot{u}$ dependence is that following an initial transient the ultimate amplitude of the oscillation has a fixed value which is independent of the initial conditions.

6.2 Averaging Method

An approximate solution to the problem of nonlinear oscillations can be obtained by a rather simple averaging technique. The principle is most clearly illustrated for a second-order system, which is assumed to be arranged into the form

$$\ddot{u} + \omega_0^2 u = \varepsilon \psi(u, \dot{u})$$
$$\psi(0, 0) = 0 \tag{6-9}$$

ε is a small positive parameter and ψ contains no linear term in u. Derivatives of ψ *evaluated at* $u = \dot{u} = 0$ are written as

$$\psi_u = \frac{\partial \psi}{\partial u}, \qquad \psi_v = \frac{\partial \psi}{\partial \dot{u}}, \qquad \psi_{uv} = \frac{\partial^2 \psi}{\partial u \, \partial \dot{u}}, \qquad \text{etc.}$$

Then $\psi_u = 0$. The steady state $u = \dot{u} = 0$ is taken to be unstable so that the oscillations will occur. From a linear stability analysis it follows that the steady state will be unstable if $\psi_v > 0$.

When $\varepsilon = 0$ the solution to Eq. (6-9) is given by Eq. (6-2). We can take the origin of the time scale so that the phase angle is zero and write

$$u(t) = \mathscr{A} \cos \omega_0 t \tag{6-10}$$

The nonlinearity would not be expected to alter the structure of the solution greatly, but we know that the following effects will occur:

1. The oscillations will not be sinusoidal and might have a nonzero mean value. If we think in terms of a Fourier series representation of the solution, we then expect not only the fundamental mode, as in Eq. (6-10), but all its harmonics as well.
2. The frequency of the oscillations will be altered by the presence of the nonlinearity and will differ from ω_0.

The essence of the approximating procedure is to assume that as long as ε is small the higher harmonics are dominated by the fundamental, the amplitude is small, and the frequency ω is not significantly different from ω_0. We therefore seek a solution in the form

$$u(t) \simeq \mathscr{A}_0 + \mathscr{A}_1 \cos \omega t \qquad (6.11a)$$

$$\dot{u}(t) \simeq -\omega \mathscr{A}_1 \sin \omega t \qquad (6.11b)$$

and estimate the values of $\mathscr{A}_0$ and $\mathscr{A}_1$.

We first establish the relation between $\mathscr{A}_0$ and $\mathscr{A}_1$ by averaging Eq. (6-9) over one period:

$$\frac{\omega}{2\pi} \int_0^{2\pi/\omega} [\ddot{u} + \omega_0^2 u] \, dt = \frac{\varepsilon\omega}{2\pi} \int_0^{2\pi/\omega} \psi(u, \dot{u}) \, dt \qquad (6\text{-}12)$$

The first term in the integral on the left integrates to zero, while the second, using Eq. (6-11a), integrates to $\omega_0^2 \mathscr{A}_0$. To obtain the integral on the right we expand ψ through third order in a Taylor series,

$$\psi(u, \dot{u}) \simeq \psi_v \dot{u} + \tfrac{1}{2}\psi_{uu} u^2 + \psi_{uv} u\dot{u} + \tfrac{1}{2}\psi_{vv} \dot{u}^2 + \tfrac{1}{6}\psi_{uuu} u^3$$

$$+ \tfrac{1}{2}\psi_{uuv} u^2 \dot{u} + \tfrac{1}{2}\psi_{uvv} u\dot{u}^2 + \tfrac{1}{6}\psi_{vvv} \dot{u}^3 \qquad (6\text{-}13)$$

Substituting Eqs. (6-11) for u and $\dot{u}$ the integration on the right of Eq. (6-12) can be carried out. Most terms integrate to zero, and we obtain

$$\omega_0^2 \mathscr{A}_0 = \frac{\varepsilon}{4} \{ 2\mathscr{A}_0^2 \psi_{uu} + \mathscr{A}_1^2 [\psi_{uu} + \omega^2 \psi_{vv} + \mathscr{A}_0 \{\psi_{uuu} + \omega^2 \psi_{uvv}\}] \} \qquad (6\text{-}14)$$

To first order in ε the solution for $\mathscr{A}_0$ in terms of $\mathscr{A}_1$ is

$$\mathscr{A}_0 = \frac{\varepsilon\mathscr{A}_1^2}{4\omega_0^2} [\psi_{uu} + \omega^2 \psi_{vv}] \qquad (6\text{-}15)$$

Note that $\mathscr{A}_0$ is of order ε relative to $\mathscr{A}_1$. Interestingly, this relation, so easily obtained, provides the solution to a practical question alluded to in Sec. 1.2. We noted there that an oscillating system might have an average output which was better than one forced to operate at steady state. Conditions under which this will be true are defined entirely by the algebraic sign of $\psi_{uu} + \omega^2 \psi_{vv}$, where ω^2 is approximately equal to ω_0^2.

To obtain an equation for $\mathscr{A}_1$ we return to the energy integral in the preceding section. Substituting Eqs. (6-11) for u and $\dot{u}$ and the expansion

(6-13) into Eq. (6-8) and carrying out the integrations (most of which are zero) gives

$$\mathscr{A}_1^2\omega\psi_v + \mathscr{A}_0\mathscr{A}_1^2\omega\psi_{uv} + \tfrac{1}{2}\mathscr{A}_1^2\omega\psi_{uuv}[\mathscr{A}_0^2 + \tfrac{1}{4}\mathscr{A}_1^2] + \tfrac{1}{8}\mathscr{A}_1^3\omega^3\psi_{vvv} = 0 \quad (6\text{-}16)$$

For simplicity we shall neglect the $\mathscr{A}_0$ terms and obtain a solution which is zero order in ε. Together with Eq. (6-15) this gives a solution for $\mathscr{A}_1$ as

$$\mathscr{A}_1 = \left[-\frac{8\psi_v}{\psi_{uuv} + \omega^2\psi_{vvv}}\right]^{1/2} \quad (6\text{-}17)$$

$\mathscr{A}_1$ must be real. Since $\psi_v > 0$, it follows that a necessary condition for sustained oscillations is $\psi_{uuv} + \omega^2\psi_{vvv} < 0$, where we substitute ω_0 for ω to first approximation. It is worth noting that to zero order in ε the amplitude of the oscillation is independent of frequency and of ε, except insofar as ε enters into the normalization of the equation and hence into the values of ψ_{uuv} and ψ_{vvv}. We emphasize again the remarkable fact that, because of the slight nonlinearity involving $\dot{u}$, only a single amplitude can ultimately be obtained in the system, independent of initial conditions. This type of sustained oscillation is known as a *limit cycle*.

The entire analysis can be repeated without making the assumption of small ε. In that case, however, there is no guarantee that the oscillations will be close to the linear oscillations, and the neglect of the changed frequency and higher harmonics could lead to serious error.

6.3 Applications of Averaging

The most commonly used example for the study of nonlinear oscillations is the *Van der Pol equation*,

$$\ddot{u} + u = \varepsilon[1 - u^2]\dot{u} \quad (6\text{-}18)$$

which describes a linear oscillating electrical circuit with resistance coupled inductively to a triode. In this case the cubic expansion for ψ in Eq. (6-13) is exact and does not introduce a small amplitude assumption. The only nonzero derivatives are

$$\psi_v = 1 \qquad \psi_{uuv} = -2$$

Then from Eq. (6-15), $\mathscr{A}_0 = 0$, and from Eq. (6-17), $\mathscr{A}_1 = 2$. The approximate solution is

$$u(t) \simeq 2\cos t$$

which is independent of ε. Figure 6.3 shows the closed limit cycle contour in the u-$\dot{u}$ phase plane for various values of ε. For small ε the circle of radius 2 is a good approximation. For larger ε the higher harmonics cause severe distortion, which is unaccounted for in the approximation.

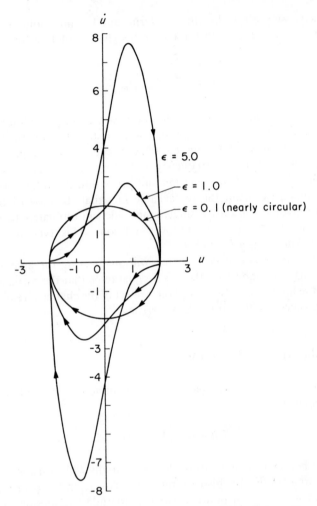

Figure 6.3 Limit cycles of the Van der Pol equation for various values of ε.

The equations for the stirred tank chemical reactor, Eqs. (5-11) and (5-12), are not in the form required for application of the theory in the preceding section, but they can be easily transformed. Equation (5-9) is differentiated once to form a second-order equation in $\bar{\eta}$, from which ξ and $\dot{\xi}$ are eliminated by use of Eqs. (5-8) and (5-9). The result is

$$\ddot{\eta} + \left\{1 + \beta_4 + \beta_1 \exp\left(\frac{\beta_3 \eta}{y_s + \eta}\right) - \frac{\beta_3 y_s^2}{[y_s + \eta]^2}\left[\beta_5 + \frac{\beta_4 \eta}{y_s}\right]\right\}\dot{\eta}$$

$$+ \left[1 + \beta_1 \exp\left(\frac{\beta_3 \eta}{y_s + \eta}\right)\right]\left[\beta_4 \eta + \beta_5 y_s\left[1 - \exp\left(\frac{\beta_3 \eta}{y_s + \eta}\right)\right]\right] - \frac{\beta_3 y_s}{[y_s + \eta]^2}\dot{\eta}^2 = 0$$

$$(6\text{-}19)$$

$$\beta_1 = \alpha e^{-\gamma/y_s} \qquad \beta_3 = \frac{\gamma}{y_s}$$

$$\beta_4 = 1 + \delta \qquad \beta_5 = \frac{\alpha F(y_s)[1 + \delta]}{y_s}$$

η is the deviation from the steady-state temperature. The β notation is adopted to facilitate comparison with the work of J. M. Douglas and N. Y. Gaitonde. The linearized form of Eq. (6-19) is

$$\ddot{\eta} - \varepsilon\dot{\eta} + \omega_0^2 \eta = 0 \qquad (6\text{-}20)$$

$$\varepsilon = -1 - \beta_1 - \beta_4 + \beta_3 \beta_5 \qquad (6\text{-}21\text{a})$$

$$\omega_0^2 = [1 + \beta_1]\beta_4 - \beta_3 \beta_5 \qquad (6\text{-}21\text{b})$$

$\omega_0^2 > 0$ corresponds to the slope condition, Eq. (4-8). $\varepsilon > 0$ corresponds to violation of the second necessary condition for stability, Eq. (4-9), so that the steady state cannot be attained, and the system can oscillate.

ε is assumed to be small, and Eq. (6-19) is rearranged into the form of Eq. (6-9). Direct application of the results of the preceding section then leads to the result

$$\mathscr{A}_0 = \frac{\mathscr{A}_1^2 \gamma[1 + \delta]}{4 y_s^4 \omega_0^2}\{\phi\gamma + y_s[2\phi - \gamma]\} \qquad (6\text{-}22)$$

$$\mathscr{A}_1 = 2y_s^2 \left\{\frac{\varepsilon}{\gamma}\frac{1}{[1 + \delta][3\phi - y_s] - \alpha e^{-\gamma/y_s}[y_s - \frac{1}{2}\gamma]}\right\}^{1/2} \qquad (6\text{-}23)$$

Two important results follow. Since $\mathscr{A}_1$ must be real in order for oscillations to occur, a necessary condition for a limit cycle is

$$[1 + \delta][3\phi - y_s] - \alpha e^{-\gamma/y_s}[y_s - \tfrac{1}{2}\gamma] > 0 \qquad (6\text{-}24)$$

Positive $\mathscr{A}_0$ indicates a mean temperature which exceeds the steady state, with a corresponding mean conversion in excess of the steady state. The condition under which oscillations will improve performance is then

$$\phi\gamma + y_s[2\phi - \gamma] > 0 \qquad (6\text{-}25)$$

6.4 Averaging in Higher-Order Systems

The averaging procedure extends easily to higher-order systems. We make use of the fact that if the linear system

$$\dot{\xi}_i = \sum_j A_{ij}\xi_j \tag{6-26}$$

has a periodic solution, the solution has the form

$$\xi_i(t) = \mathscr{A}\lambda_i \cos(\omega_0 t + \phi_i) \tag{6-27}$$

λ_i and ϕ_i are constants, and the matrix $\mathbf{A}$ has the property that

$$\int_0^{2\pi/\omega_0} \sum_{i,j} \lambda_i \lambda_j \cos(\omega_0 t + \phi_i) A_{ij} \cos(\omega_0 t + \phi_j)\, dt = 0 \tag{6-28}$$

We consider nonlinear systems of the form

$$\dot{\xi}_i = \sum_j A_{ij}\xi_j + \varepsilon b_i(\boldsymbol{\xi}) \tag{6-29}$$

where ε is a small parameter and Eq. (6-27) is the solution for $\varepsilon \to 0$. We then seek a solution in the form

$$\xi_i = \mathscr{A}_{0i} + \mathscr{A}_1 \lambda_i \cos(\omega_0 t + \phi_i) \tag{6-30}$$

where we have set the frequency equal to ω_0. $b_i(\boldsymbol{\xi})$ is expanded through third order as

$$b_i(\boldsymbol{\xi}) = \sum_j \alpha_{ij}\xi_j + \sum_{j,k} \beta_{ijk}\xi_j\xi_k + \sum_{j,k,l} \gamma_{ijkl}\xi_j\xi_k\xi_l \tag{6-31}$$

Equations (6.30) and (6-31) are substituted into Eq. (6-29), and the latter equation is integrated over one period, leading immediately to the linear equations

$$\sum_j A_{ij}\mathscr{A}_{0j} = -\tfrac{1}{2}\varepsilon\mathscr{A}_1^2 \sum_{j,k} \beta_{ijk}\lambda_j\lambda_k \cos(\phi_j - \phi_k) \tag{6-32}$$

which can be solved for the steady-state offsets $\mathscr{A}_{0j}$. Note that $\mathscr{A}_{0j}$ are of order ε relative to $\mathscr{A}_1$. To obtain an equation for $\mathscr{A}_1$ we multiply Eq. (6-29) by ξ_i and sum over i:

$$\sum_i \xi_i\dot{\xi}_i = \frac{1}{2}\frac{d}{dt}\sum_i \xi_i^2 = \sum_{i,j} \xi_i A_{ij}\xi_j + \varepsilon \sum_i \xi_i b_i(\boldsymbol{\xi}) \tag{6-33}$$

In substituting Eq. (6-30) into Eq. (6-33) we neglect the $\mathscr{A}_{0i}$ term for simplicity, giving a result accurate through zero order in ε. Integrating Eq. (6-33) over one period and using Eq. (6-28) then gives the result

$$0 = \frac{1}{2} \sum_{i,j} \lambda_i \lambda_j \alpha_{ij} \cos (\phi_i - \phi_j) + \frac{1}{8} \mathscr{A}_1^2 \sum_{i,j,k,l} \lambda_i \lambda_j \lambda_k \lambda_l \gamma_{ijkl} \mathscr{F}(\phi_i, \phi_j, \phi_k, \phi_l)$$

(6-34a)

$$\mathscr{A}_1 = 2 \left[- \frac{\displaystyle\sum_{i,j} \lambda_i \lambda_j \alpha_{ij} \cos (\phi_i - \phi_j)}{\displaystyle\sum_{i,j,k,l} \lambda_i \lambda_j \lambda_k \lambda_l \gamma_{ijkl} \mathscr{F}(\phi_i, \phi_j, \phi_k, \phi_l)} \right]^{1/2}$$

(6-34b)

$$\mathscr{F}(\phi_i, \phi_j, \phi_k, \phi_l) = 3 \cos \phi_i \cos \phi_j \cos \phi_k \cos \phi_l$$
$$+ 3 \sin \phi_i \sin \phi_j \sin \phi_k \sin \phi_l$$
$$+ \cos \phi_i \cos \phi_j \sin \phi_k \sin \phi_l + \sin \phi_i \sin \phi_j \cos \phi_k \cos \phi_l$$
$$+ \sin (\phi_i + \phi_j) \sin (\phi_k + \phi_l)$$

(6-35)

It is readily established for the system

$$\dot{\xi}_1 = \xi_2$$
$$\dot{\xi}_2 = -\omega_0^2 \xi_1 + \varepsilon \psi(\xi_1, \xi_2)$$

(6-36)

which is equivalent to Eq. (6-9), that the results of this section reduce to Eqs. (6-15) and (6-17). Here $\phi_1 = 0$, $\phi_2 = \pi/2$, $\lambda_1 = 1$, $\lambda_2 = \omega_0$, and all $\gamma_{1jkl} = 0$.

6.5 Poincaré-Lindstedt Perturbation Method

The presence of a small parameter in the nonlinear differential equation naturally suggests a perturbation solution. Perturbation solutions can be applied to nonlinear periodic systems, but great care must be taken. This point is most easily illustrated by the nonlinear equation

$$\ddot{u} + u + \varepsilon u^3 = 0$$

(6-37)

If we seek a solution in the form

$$u(t) = u_0(t) + \varepsilon u_1(t) + \varepsilon^2 u_2(t) + \cdots$$

(6-38)

substitute into Eq. (6-37) and collect terms of the same order in ε, we obtain in the usual manner a sequence of linear equations:

$$\varepsilon^0: \quad \ddot{u}_0 + u_0 = 0 \tag{6-39a}$$

$$\varepsilon^1: \quad \ddot{u}_1 + u_1 = -u_0^3 \tag{6-39b}$$

etc.

The solution to the first equation is simply the linear oscillator, $u_0 = \mathscr{A}\cos t$. Equation (6-39b) then becomes

$$\ddot{u}_1 + u_1 = -\mathscr{A}^3 \cos^3 t = -\mathscr{A}^3[\tfrac{1}{4}\cos 3t + \tfrac{3}{4}\cos t] \tag{6-40}$$

with the solution

$$u_1(t) = \frac{\mathscr{A}^3}{32}\cos 3t - \frac{3\mathscr{A}^3}{8}\, t \sin t \tag{6-41}$$

The $t \sin t$ term results from the $\cos t$ on the right of Eq. (6-40). $t \sin t$ is not periodic and grows without bound. This is known as a *secular term*. Thus, the elementary perturbation approach will not work.

The difficulty in the perturbation approach is that it neglects the small change in the frequency of the oscillation because of the nonlinearity. The Poincaré-Lindstedt method takes this change into account by "stretching" the time scale,

$$t = \tau[1 + b_1\varepsilon + b_2\varepsilon^2 + \cdots] \tag{6-42}$$

If we let a prime denote differentiation with respect to τ, Eq. (6-37) can be written as

$$u'' + [1 + b_1\varepsilon + b_2\varepsilon^2 + \cdots]^2 u + \varepsilon u^3 = 0 \tag{6-43}$$

If we now write a perturbation series as

$$u = u_0(\tau) + \varepsilon u_1(\tau) + \cdots \tag{6-44}$$

and collect terms of equal order of ε, we obtain the following series of equations:

$$\varepsilon^0: \quad u_0'' + u_0 = 0 \tag{6-45a}$$

$$\varepsilon^1: \quad u_1'' + u_1 = -2b_1 u_0 - u_0^3 \tag{6-45b}$$

etc.

$u_0(\tau)$ is simply $\mathscr{A}\cos\tau$. Then Eq. (6-45b) becomes

$$u_1'' + u_1 = -2b_1\mathscr{A}\cos\tau - \mathscr{A}^3\cos^3\tau$$

$$= -\frac{\mathscr{A}^3}{4}\cos 3\tau - \left(\frac{3\mathscr{A}^3}{4} + 2b_1\mathscr{A}\right)\cos\tau \qquad (6\text{-}46)$$

The secular term is eliminated by choosing b_1 so that the $\cos\tau$ term on the right disappears,

$$b_1 = \frac{3\mathscr{A}^2}{8}, \qquad \tau = \frac{t}{1 - (3\varepsilon\mathscr{A}^2/8)} \qquad (6\text{-}47)$$

The solution is then

$$u(t) = \mathscr{A}\cos\left(\frac{t}{1 - (3\varepsilon\mathscr{A}^2/8)}\right) + \frac{\varepsilon\mathscr{A}^3}{32}\cos\left(\frac{3t}{1 - (3\varepsilon\mathscr{A}^2/8)}\right) + \cdots \qquad (6\text{-}48)$$

For this system we have already seen in Sec. 6.1 that the amplitude is determined by the initial condition.

It is helpful to study the application of the perturbation method to the Van der Pol equation (6-18). Proceeding as above we obtain the perturbation equations

$$u_0'' + u_0 = 0 \qquad u_0 = \mathscr{A}\cos\tau \qquad (6\text{-}49a)$$

$$u_1'' + u_1 = -2b_1\mathscr{A}\cos\tau - (1 - \mathscr{A}^2\cos^2\tau)\sin\tau$$

$$= -2b_1\mathscr{A}\cos\tau - \left(1 - \frac{\mathscr{A}^2}{4}\right)\sin\tau + \frac{\mathscr{A}^2}{4}\sin 3\tau \qquad (6\text{-}49b)$$

Secular terms will occur unless the $\cos\tau$ and $\sin\tau$ terms on the right of Eq. (6-49b) are removed. Setting their coefficients to zero gives

$$b_1 = 0 \qquad \mathscr{A} = 2$$

That is, to first order the frequency is unchanged, and the solution is

$$u(t) = 2\cos t - \frac{\varepsilon}{4}\sin 3t + \cdots \qquad (6\text{-}50)$$

This agrees with the solution obtained by the averaging method through zero order in ε.

When the Poincaré-Lindstedt method is applied to the chemical reactor, Eq. (6-19), both the amplitude and frequency are computed after tedious algebra as follows:

$$\eta = \mathscr{A}_1 \cos \omega t \qquad (6\text{-}51)$$

where $\mathscr{A}_1$ is the same as the result computed by the averaging method, Eq. (6-23), and

$$\omega = \omega_0 \left[1 - \frac{\beta_3 (\beta_1\beta_4 + \frac{5}{2}\beta_3\beta_5 + 3\beta_5 + \frac{3}{2}\beta_1\beta_3\beta_4 - 2\beta_1)}{8\omega_0^2} \mathscr{A}_1^2 \right] \qquad (6\text{-}52)$$

ω_0 is given by Eq. (6-21b). When the second perturbation (ε^2) equation is considered, terms of order $\mathscr{A}_1^2$ are obtained, including higher harmonics and a constant term. The constant term is the same as the result from the averaging method, Eq. (6-22), to order $\mathscr{A}_1^2$.

6.6 Fourier Methods

The most extensive information about the behavior of a nonlinear oscillating process is obtained by direct Fourier methods. While this is also the most tedious approach by far, the reward is information about the transient approach to the oscillations as well as ultimate amplitude and phase angles. In addition, nonlinear stability bounds like those of the Eckhaus method (Sec. 5.6) are obtained for some systems. A number of very similar approaches are used. We shall describe here the *cascade method*, which has been applied to several important problems in nonlinear distributed parameter systems and to which we shall return later. It is closely related to the more common *method of Krylov and Bogoliubov*.

The important observation for all Fourier methods is that the parameters ω_0 and ε in Eq. (6-9) play different roles. ω_0 defines the time scale for oscillations, while ε defines the time scale for growth or decay of the amplitude of oscillations. If $\varepsilon \ll \omega_0$, then the amplitude and phase angle will change very little over one period, and we may treat the two time scales differently. Thus, we shall be looking for solutions with the general behavior

$$u(t) = \mathscr{A} \cos (\omega_0 t + \phi)$$

where $\mathscr{A}$ and ϕ are slowly changing functions of time. The phase change can equivalently be considered to depend on the amplitude, $\mathscr{A}$. When $\mathscr{A}$ is small, we expect it to grow or decay according to the linearized version of

Eq. (6-9), while for long times we expect $\mathscr{A}$ and ϕ to reach asymptotic values characteristic of the limit cycle.

It is convenient to use the exponential representation of a Fourier series for a periodic function,

$$u(t) = \sum_k [C_k e^{ik\Omega t} + C_k^* e^{-ik\Omega t}] \tag{6-53}$$

Recall that

$$e^{ik\Omega t} = \cos k\Omega t + i \sin k\Omega t$$

C_k is a complex coefficient,

$$C_k = C_{Rk} + iC_{Ik}$$

and C_k^* is the complex conjugate

$$C_k^* = C_{Rk} - iC_{Ik}$$

It is easily verified that Eq. (6-53) is equivalent to the more familiar expansion in sines and cosines. Both the coefficients C_k and frequency Ω will depend on the amplitude, $\mathscr{A}$, and Ω must approach ω_0 for small $\mathscr{A}$. Equation (6-53) is substituted directly into the nonlinear differential equation (with nonlinearities usually expanded through third order), and terms are grouped according to the number of the harmonic ($e^{ik\Omega t}$). An infinite set of equations is obtained by equating the harmonic coefficients on each side of Eq. (6-53). The equations will include terms $[\partial C_k/\partial \mathscr{A}]\mathscr{A}$, $[\partial^2 C_k/\partial \mathscr{A}^2]\mathscr{A}^2$, and $d\Omega t/dt$. It should be noted that we can always choose the origin of the time axis so that the phase angle of the fundamental frequency ($k = 1$) is zero, which is equivalent to having C_1 real. This is not true for $k > 1$, however.

The amplitude dependence of the Fourier coefficients is assumed to be a power series

$$C_k = \sum_{n \geq k} 2^{-n} \phi_{kn} \mathscr{A}^n \tag{6-54}$$

All ϕ_{1n} are real, and the normalization is established by taking $\phi_{11} = 1$. The logic for taking the sum $n \geq k$ is as follows. The kth harmonic is assumed to be of order $\mathscr{A}^k$. The nonlinear interaction of harmonic terms will lead to a lower harmonic; e.g.,

$$\mathscr{A}^p e^{ip\Omega t} \mathscr{A}^q e^{-iq\Omega t} = \mathscr{A}^{p+q} e^{i(p-q)\Omega t}$$

Thus, a given harmonic coefficient will also have terms containing higher powers of $\mathscr{A}$. The amplitude and frequency are assumed to evolve according to ordinary differential equations:

$$\dot{\mathscr{A}} = a_0 \mathscr{A} + \frac{1}{2} a_1 \mathscr{A}^2 + \frac{1}{4} a_2 \mathscr{A}^3 + \cdots \tag{6-55}$$

$$\frac{d}{dt} \Omega t = b_0 + \frac{1}{2} b_1 \mathscr{A} + \frac{1}{4} b_2 \mathscr{A}^2 + \cdots \tag{6-56}$$

[The factors of 2 in Eqs. (6-54) to (6-56) are introduced to make $\mathscr{A}$ correspond to the amplitude used previously.] Equations (6-54) through (6-56) are substituted into the equation for each harmonic coefficient after which, in each equation, coefficients of like powers of $\mathscr{A}$ are equated. The result is a doubly infinite set of equations for the coefficients ϕ_{kn}, $\{a_i\}$, and $\{b_i\}$, Clearly, a_0 and b_0 must correspond to the behavior of the linearized system.

The process described above is extremely tedious, though a summation convention introduced by W. C. Reynolds and M. C. Potter helps somewhat, and we might expect that in the future the bookkeeping of multiplying and sorting Fourier series will be done by a computer. We shall sketch out some of the steps for the Van der Pol equation (6-18). The linearized version,

$$\ddot{u} - \varepsilon \dot{u} + u = 0 \tag{6-57}$$

has a solution

$$u = \mathscr{A}(0) e^{\varepsilon t/2} \cos \sqrt{\frac{1 - \varepsilon^2}{4}} t \tag{6-58}$$

where $\mathscr{A}(0)$ is the initial amplitude. We therefore expect $a_0 = \varepsilon/2$, $b_0 = [1 - \varepsilon^2/4]^{1/2}$.

We tabulate below the coefficient equations up to $n = 2$ according to the Fourier component (k) and the power of $\mathscr{A}(n)$, making use of the facts that $\phi_{11} = 1$ and that all ϕ_{1n} are real:

$k = 0, n = 0:$ $\phi_{00} = 0$

$k = 0, n = 1:$ $\phi_{01}[a_0^2 + 1 - \varepsilon a_0] = 0$

$k = 0, n = 2:$ $\phi_{01}[3a_0 a_1 - a_1 \varepsilon] + \phi_{02}[4a_0^2 + 1 - 2a_0 \varepsilon] = 0$

$k = 1, n = 1:$ $a_0^2 + 2ia_0 b_0 - b_0^2 + 1 - \varepsilon a_0 - i\varepsilon b_0 = 0$

$k = 1, n = 2:$ $3a_0 a_1 + 3ia_0 b_1 + 2ia_1 b_0 - 2b_1 b_0 - \varepsilon a_1 - i\varepsilon b_1$
$\qquad\qquad\quad + \phi_{12}[4a_0^2 + 4ia_0 b_0 - b_0^2 + 1 - 4a_0 \varepsilon - i\varepsilon b_0] = 0$

$k = 2, n = 2:$ $\phi_{22}[4a_0^2 + 8ia_0 b_0 - 4b_0^2 + 1 - 2a_0 \varepsilon - 2i\varepsilon b_0] = 0$

From the real and imaginary parts of $k = 1$, $n = 1$ it follows at once that

$$a_0 = \frac{\varepsilon}{2} \qquad b_0 = \sqrt{\frac{1 - \varepsilon^2}{4}}$$

as expected from the linear solution. Then from $k = 0$, $n = 0$; $k = 0$, $n = 1$; $k = 0$, $n = 2$; and $k = 2$, $n = 2$ we obtain

$$\phi_{00} = \phi_{01} = \phi_{02} = \phi_{22} = 0$$

Thus, through at least second order in amplitude there is no constant term representing a mean response, and the contribution of the second harmonic is at least third order. Finally, it can be shown that the real and imaginary parts of $k = 1$, $n = 2$ are not compatible with the above results unless

$$a_1 = b_1 = \phi_{12} = 0$$

Next we tabulate the equations for $n = 3$, making use of the results which we have already obtained to shorten the equations:

$k = 0$, $n = 3$: $\quad \phi_{03}[9a_0^2 + 1 - 3a_1\varepsilon] = 0$

$k = 1$, $n = 3$: $\quad 4a_1 a_3 + 2ia_2 b_0 + 4ia_0 b_2 - 2b_0 b_2 - a_2 \varepsilon - i\varepsilon b_2 + 3a_0 \varepsilon + i\varepsilon b_0$
$\qquad\qquad\qquad + \phi_{13}[9a_0^2 + 6ia_0 b_0 - b_0^2 + 1 - 3a_0 \varepsilon - i\varepsilon b_0] = 0$

$k = 2$, $n = 3$: $\quad \phi_{23}[9a_0^2 + 12ia_0 b_0 - 4b_0^2 + 1 - 3a_0 \varepsilon - 2ib_0 \varepsilon] = 0$

$k = 3$, $n = 3$: $\quad a_0 \varepsilon + i\varepsilon b_0 + \phi_{33}[9a_0^2 + 18ia_0 b_0 - 9b_0^2 + 1 - 3a_0 \varepsilon - 3i\varepsilon b_0]$
$\qquad\qquad\qquad\qquad\qquad\qquad\qquad\qquad\qquad\qquad\qquad\qquad\qquad = 0$

From $k = 0$, 2, and 3 we obtain

$$\phi_{03} = \phi_{23} = 0$$

$$\phi_{33} = \frac{7\varepsilon^2}{32} + i\frac{\varepsilon}{8}$$

We have neglected terms in ε^3 and higher in ϕ_{33} because of the requirement $\varepsilon \ll 1$. Finally, from the real and imaginary parts of $k = 1$, $n = 3$ we obtain

$$b_2 = \frac{\varepsilon^2 b_0}{2}$$

and the equation

$$2a_2 + \frac{\varepsilon^3}{2} + \varepsilon + 2\varepsilon\phi_{13} = 0 \tag{6-59}$$

This seeming ambiguity is resolved by noting that as $\varepsilon \to 0$ we must reduce to Eq. (6-58) for $\varepsilon = 0$, in which the amplitude enters linearly. Thus, ϕ_{13} must be at least of order ε, so that the term $\phi_{13}\mathscr{A}^3$ will vanish as $\varepsilon \to 0$. In that case we can write Eq. (6-59) as

$$2a_2 + \varepsilon + \cdots = 0$$

$$a_2 = -\frac{\varepsilon}{2}$$

Equation (6-55) for the amplitude is now

$$\dot{\mathscr{A}} = \frac{\varepsilon}{2}\mathscr{A} - \frac{\varepsilon}{8}\mathscr{A}^3 \tag{6-60}$$

which has a solution

$$\mathscr{A}(t) = \frac{2e^{\varepsilon t/2}}{\{[4/\mathscr{A}^2(0)] - 1 + e^{\varepsilon t}\}^{1/2}} \tag{6-61}$$

Thus, for small εt and small $\mathscr{A}$ the behavior is initially like the linear solution, Eq. (6-58). Eventually, however, the amplitude grows to an equilibrium value of $\mathscr{A} = 2$, and we reach a limit cycle with a frequency

$$t \to \infty: \quad \Omega = b_0 + \frac{b_2\mathscr{A}^2}{4} + \cdots = 1 + \cdots$$

where any neglected terms are of order higher than ε^2.

It is interesting to note that we have not necessarily assumed that $\varepsilon > 0$ for this analysis. When $\varepsilon < 0$ we know from linear theory that the steady state is stable to infinitesimal disturbances. According to Eq. (6-61), the nonlinear behavior is the same as long as $\mathscr{A}(0)$ is small. In fact, $\mathscr{A}(t)$ will go to zero as long as $\mathscr{A}(0) < 2$. However, it is readily established that when $\mathscr{A}(0) > 2$ and $\varepsilon < 0$ the denominator in Eq. (6-61) will go to zero and $\mathscr{A}(t)$ will grow without bound. Thus, we find that within the accuracy of the approximation there is a critical amplitude for the system. Disturbances smaller than this amplitude will decay, but the steady state is unstable to larger disturbances. This observation is in agreement with the known behavior of the Van der Pol equation for $\varepsilon < 0$. Note the similarity between this approach and the Eckhaus method (Sec. 5.6).

When the cascade method is applied to the chemical reactor equation (6-15) we obtain the solution

$$\eta(t) = \mathscr{A}_0 + \mathscr{A} \cos \omega t + \cdots$$

where $\mathscr{A}_0$ is given by Eq. (6-22), ω by Eq. (6-52), and

$$\mathscr{A}(t) = \frac{\mathscr{A}_1}{(1 - \{1 - [\mathscr{A}_1^2/\mathscr{A}^2(0)]\}e^{-\varepsilon t})^{1/2}}$$

$\mathscr{A}_1$ is given by the result for the averaging method, Eq. (6-23).

BIBLIOGRAPHICAL NOTES

The averaging method is developed in

Denn, M. M., and J. R. Black, *Chem. Eng. Sci.*, *28*, 1515 (1973).

The Poincaré-Lindstedt, Krylov-Bogoliubov, and other classical nonlinear methods are covered in many texts, such as

Bellman, R., *Perturbation Techniques in Mathematics, Physics, and Engineering*, Holt, Rinehart and Winston, New York, 1964.

Davis, H. T., *Introduction to Nonlinear Differential and Integral Equations*, Dover, New York, 1962.

Douglas, J. M., *Process Dynamics and Control*, Prentice-Hall, Englewood Cliffs, N.J., 1972.

Lapidus, L., and R. Luus, *Optimal Control of Engineering Processes*, Ginn/Blaisdell, Waltham, Mass., 1967.

Minorsky, N., *Nonlinear Oscillations*, Van Nostrand, Rinehold, New York, 1962.

The summation convention for manipulating terms in the cascade method is in

Reynolds, W. C., and M. C. Potter, *J. Fluid Mech.*, *27*, 465 (1967).

Applications of several different nonlinear methods to the stirred tank reactor are in the books by Douglas and Luus and Lapidus, cited above. See also,

Beek, J., *AIChE J.*, *18*, 228 (1972).

Douglas, J. M., and N. Y. Gaitonde, *Ind. Eng. Chem. Fundamentals*, *6*, 26, (1967).

Douglas's book contains other applications and references, including the use of a positive feedback control system to enhance oscillations for improved time-averaged system performance. Literature in this area is also tabulated in

Denn, M. M., in V. W. Weekman, Jr., ed., *Annual Review of Industrial and Engineering Chemistry, 1970*, American Chemical Society, Washington, D.C., 1972.

For an application to the closely related problem of oscillations in a continuous crystallizer, see

Yu, K. M., and J. M. Douglas, to be published; K. M. Yu, Ph.D. Dissertation, University of Massachusetts, Amherst, 1972.

The possibility of sustained oscillations is of particular interest in biological systems, and considerable effort has been devoted to the development of kinetic schemes which allow oscillation. See

Glansdorff, P., and I. Prigogine, *Thermodynamic Theory of Structure, Stability and Fluctuations*, Wiley, New York, 1972.

Higgins, J., *Ind. Eng. Chem.*, *59*, no. 5, 19 (1967).

The Lotka-Volterra kinetic scheme is considered in great detail in the book by Davis, cited above. In an interesting series of papers Horn and coworkers have developed methods for identifying kinetic schemes which cannot exhibit oscillations. See

Feinberg, M., *Arch. Rat. Mech. Anal.*, *49*, 187 (1972).

Horn, F., *Arch. Rat. Mech. Anal.*, *49*, 172 (1972).

Horn, F., and R. Jackson, *Arch. Rat. Mech. Anal.*, *47*, 81 (1972).

In a more general context, a theorem by H. Bendixson which establishes conditions that preclude oscillations is applied in the texts cited above.

Uniqueness in Distributed Parameter Systems

7

7.1 Introduction

In the remainder of the book we shall be concerned with the dynamical behavior of systems described by partial differential equations. Such *distributed parameter processes* arise when there is a spatial variation in the properties of the system. The development parallels Chapters 2 through 6. In this chapter we shall deal with the question of uniqueness of the steady state. Unlike the study of ordinary differential equations, we cannot state general theorems regarding uniqueness for partial differential equations. For special classes of equations results can be obtained, and these are detailed in textbooks on the subject. We shall consider in this chapter only the case of the equations describing reaction and heat transport in a porous catalyst.

7.2 Catalyst Particle Equations

The physicochemical problem of interest here is the chemical reaction $A \rightarrow B$ in a porous catalyst pellet. Reactant A is in the fluid (gas or liquid) surrounding the catalyst particle at concentration c_f and temperature T_f. A diffuses into the catalyst and reacts to form B on the surface of a pore. Product B then diffuses out. For simplicity we shall take the catalyst geometry as a slab of thickness $2L$ which is long in the two other spatial directions. This is shown in Fig. 7.1. We shall also assume that the reaction is first order and that there

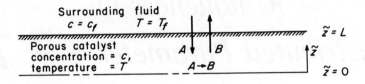

Figure 7.1 Schematic diagram of diffusion and reaction in a porous slab catalyst.

is no surface resistance to the transport of mass or energy. In that case the mass and energy balance equations are

$$\frac{\partial c}{\partial \tilde{t}} = D \frac{\partial^2 c}{\partial \tilde{z}^2} - k e^{-E/RT} c \tag{7-1}$$

$$\frac{\partial T}{\partial \tilde{t}} = D_T \frac{\partial^2 T}{\partial \tilde{z}^2} + \frac{[-\Delta H]k}{\rho c_p} e^{-E/RT} c \tag{7-2}$$

$$c = c_f, \qquad T = T_f \qquad \text{at } \tilde{z} = L \tag{7-3a}$$

$$\frac{\partial c}{\partial \tilde{z}} = 0, \qquad \frac{\partial T}{\partial \tilde{z}} = 0 \qquad \text{at } \tilde{z} = 0 \tag{7-3b}$$

D and D_T are the effective mass and thermal diffusivities in the catalyst. The other nomenclature is defined in Sec. 2.3.

The variables are made dimensionless as follows:

$$x = \frac{c}{c_f} \qquad\qquad y = \frac{T}{T_f}$$

$$\alpha = \frac{L^2 k}{D} \qquad\qquad \mathcal{L} = \frac{D}{D_T} \qquad \gamma = \frac{E}{RT_f}$$

$$\beta = \frac{[-\Delta H] c_f \mathcal{L}}{\rho c_p T_f} \qquad t = \frac{D\tilde{t}}{L^2} \qquad z = \frac{\tilde{z}}{L}$$

The ratio of mass to thermal diffusivity, $\mathcal{L}$, is sometimes known as the *Lewis number*, though the terminology is used in the literature for both this group and its reciprocal. After substituting the dimensionless variables Eqs. (7-1) through (7-3) become

$$\frac{\partial x}{\partial t} = \frac{\partial^2 x}{\partial z^2} - \alpha x e^{-\gamma/y} \tag{7-4}$$

$$\mathscr{L}\frac{\partial y}{\partial t} = \frac{\partial^2 y}{\partial z^2} + \alpha\beta x e^{-\gamma/y} \tag{7-5}$$

$$x = y = 1 \qquad \text{at } z = 1 \tag{7-6a}$$

$$\frac{\partial x}{\partial z} = \frac{\partial y}{\partial z} = 0 \qquad \text{at } z = 0 \tag{7-6b}$$

Note the similarity in structure to the equations for a stirred tank, Eqs. (2-9) and (2-10). The nomenclature has been chosen here to emphasize this similarity. In further analogy to the stirred tank we define the function

$$F(y) = [\beta + 1 - y]e^{-\gamma/y} \tag{7-7}$$

which corresponds to Eq. (2-22) for an adiabatic reactor. The properties of $F(y)$ are examined in Sec. 2-5.

7.3 Steady-State Catalyst Particle

At steady state the time derivatives are zero and Eqs. (7-4) to (7-6) become

$$x_s'' - \alpha x_s e^{-\gamma/y_s} = 0 \tag{7-8}$$

$$y_s'' + \alpha\beta x_s e^{-\gamma/y_s} = 0 \tag{7-9}$$

$$x_s(1) = y_s(1) = 1 \tag{7-10a}$$

$$x_s'(0) = y_s'(0) = 0 \tag{7-10b}$$

The subscript s denotes steady state and the prime differentiation with respect to z. These equations can be greatly simplified if we multiply Eq. (7-8) by β and add it to Eq. (7-9), yielding

$$\frac{d^2}{dz^2}[\beta x_s + y_s] = 0$$

Upon integrating twice and using the boundary conditions, we then obtain

$$\beta x_s + y_s = 1 + \beta \tag{7-11}$$

Thus, we can eliminate x_s from Eq. (7-9) and obtain the final steady state equation,

$$y_s'' + \alpha F(y_s) = 0 \tag{7-12}$$

$$y_s'(0) = 0, \qquad y_s(1) = 1 \tag{7-13}$$

$F(y)$ is defined by Eq. (7-7). Since $0 \le x_s \le 1$, a consequence of Eq. (7-11) is

$$1 \le y_s \le 1 + \beta \tag{7-14}$$

Note that the steady state is independent of the Lewis number.

The derivation of conditions for uniqueness proceeds in a manner analogous to the flow reactor. Let y_{1s} and y_{2s} be two solutions to Eqs. (7-12) and (7-13) and define

$$w = y_{1s} - y_{2s} \tag{7-15}$$

Then

$$w'' + \alpha[F(y_{1s}) - F(y_{2s})] = 0 \tag{7-16}$$

$$w'(0) = w(1) = 0 \tag{7-17}$$

From the mean value theorem we can write

$$F(y_{1s}) - F(y_{2s}) = F'(\bar{y}(z))w$$

where $\bar{y}(z)$ is some function which lies between y_{1s} and y_{2s} at each value of z. Thus, Eq. (7-16) becomes

$$w'' + \alpha F'(\bar{y})w = 0 \tag{7-18}$$

Equation (7-18) belongs to the class of equations studied in Appendix D, where we identify $\alpha F'(\bar{y}(z))$ with $\psi(z)$ in Eq. (D-1). Using inequality (D-6) we then establish that $w(z)$ must be zero for

$$\alpha \max_{1 \le y \le 1+\beta} F'(y) < \frac{\pi^2}{4} \tag{7-19}$$

If $\max F'(y) < 0$, then inequality (7-19) is trivially satisfied. This is equivalent to (compare Sec. 2-5)

$$\beta\gamma < 1 \tag{7-20}$$

A stronger condition for uniqueness can be derived provided that $\alpha \max F'(y) < \pi^2$. With this restriction we can apply the result of Sec. D.2 and observe that any solution $w(z)$ to Eq. (7-18) cannot change sign in $0 \leq z < 1$. Thus, if we have two solutions $y_{1s}(z)$ and $y_{2s}(z)$, the difference between them is always of the same sign, and one is always larger than the other. We shall take $y_{2s} > y_{1s}$, $0 \leq z < 1$. We write Eq. (7-12) for y_{1s} and y_{2s},

$$y''_{1s} + \alpha F(y_{1s}) = 0$$

$$y''_{2s} + \alpha F(y_{2s}) = 0$$

We multiply the first equation by $y_{2s} - 1$, the second by $y_{1s} - 1$, subtract, and integrate over z from zero to 1. This gives

$$\int_0^1 \{[y_{2s} - 1]F(y_{1s}) - [y_{1s} - 1]F(y_{2s})\} \, dz$$

$$= \int_0^1 \left\{ \frac{F(y_{1s})}{y_{1s}-1} - \frac{F(y_{2s})}{y_{2s}-1} \right\}[y_{1s} - 1][y_{2s} - 1] \, dz = 0 \qquad (7\text{-}21)$$

Since y_{1s} and y_{2s} can never be equal and $y_{2s} > y_{1s} \geq 1$, Eq. (7-21) can never be satisfied if $F(y)/[y - 1]$ is a monotonic function of z, or, equivalently,

$$\frac{d}{dy} \frac{F(y)}{y - 1} \leq 1, \qquad 1 \leq y \leq 1 + \beta$$

We showed in Sec. 2.5 that this is equivalent to the condition for uniqueness

$$\beta\gamma < 4 + 4\beta \qquad (7\text{-}22)$$

7.4 Concluding Remarks

The possibility of nonuniqueness of the steady state in a spatially distributed system is a very real one. Figure 7.2 shows three temperature profiles in a catalyst particle computed for $\alpha = 1.23 \times 10^7$, $\beta = 0.7$, $\gamma = 20$, where inequality (7-22) is violated. These parameters are within the possible range of processing conditions. Nonuniqueness is equally possible in other situations of interest. In the flow of a Newtonian liquid in a pipe there is a wide region in the neighborhood of Reynolds number 2100 where the flow may be either laminar or turbulent, depending on operating conditions. Useful analytical criteria for uniqueness can rarely be developed. When they can, it is usually by means of the general approach in the preceding section.

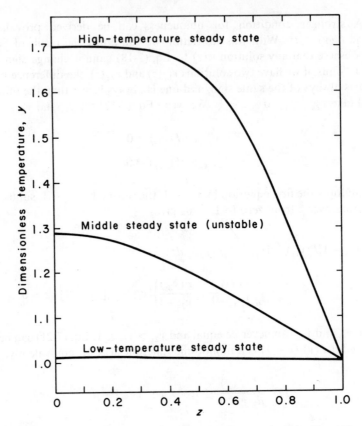

Figure 7.2 Temperature profiles in a slab catalyst. $\alpha = 1.23 \times 10^7$, $\beta = 0.7, \gamma = 20$. [After L. R. Raymond and N. R. Amundson. *AIChE J.*, *11*, 339 (1965), with permission.]

There is a very different approach to uniqueness which is sometimes useful. If it can be shown that a steady state is asymptotically stable with respect to all possible perturbations, then the steady state must be unique, since no other steady state can ever be reached. We shall return briefly to this idea following the discussion of stability to finite disturbances in Chapter 15.

BIBLIOGRAPHICAL NOTES

The subject matter of the chapter is covered completely in

Aris, R., *The Mathematical Theory of Diffusion and Reaction in Permeable Catalysts*, Oxford University Press, Inc., New York, 1974.

For the basis of the model equations, see also

Peterson, E. E., *Chemical Reaction Analysis*, Prentice-Hall, Englewood Cliffs, N.J., 1965.

The derivation of the uniqueness criterion follows

Luss, D., *Chem. Eng. Sci.*, *26*, 1713 (1971).

The result does not carry over directly to other geometries. See Aris's book and

Luss, D., *Chem. Eng. Sci.*, *27*, 2299 (1972).

Uniqueness is also considered in detail in

Gavalas, G. R., *Nonlinear Differential Equations of Chemically Reacting Systems*, Springer-Verlag New York, Inc., New York, 1968.
Perlmutter, D. D., *Stability of Chemical Reactors*, Prentice-Hall, Englewood Cliffs, N.J., 1972.

Perlmutter also deals with the similar problem of the adiabatic tubular reactor. Related problems in combustion are in

Frank-Kamenetskii, D. A., *Diffusion and Heat Transfer in Chemical Kinetics*, 2nd ed., Plenum, New York, 1969.

Galerkin's Method　　　　8

8.1 Introduction

It will frequently be necessary in the chapters which follow to obtain the eigenvalues for linear homogeneous differential equations. In a very few cases the eigenvalues may be obtained exactly by analytical means, but it is usually necessary to employ approximation methods. We shall discuss here the most common of these approximation techniques, *Galerkin's method*. The foundations of the method lie in the calculus of variations, and when the eigenvalue problem is derived from a minimization problem, Galerkin's method is equivalent to the classical Rayleigh-Ritz method.

For illustrative purposes in this chapter we shall use the system

$$x^{IV} - \lambda x = 0 \qquad (8\text{-}1)$$

$$x'(0) = x'''(0) = x(1) = x''(1) = 0$$

Superscripts refer to differentiation with respect to the independent variable, z. Equation (8-1) always has a trivial solution $x(z) = 0$. It is readily established that nontrivial solutions exist if and only if λ takes on certain values known as *eigenvalues*,

$$\lambda_n = \left[\frac{2n-1}{2}\right]^4 \pi^4, \qquad n = 1, 2, \ldots \qquad (8\text{-}2)$$

in which case the corresponding solutions (*eigenfunctions*) are

$$x_n(z) = \mathscr{A} \cos \lambda_n^{1/4} z \qquad (8\text{-}3)$$

Equation (8-1) is equivalent to the coupled system

$$x'' - y = 0$$

$$y'' - \lambda x = 0 \qquad (8\text{-}4)$$

$$x'(0) = y'(0) = x(1) = y(1) = 0$$

The latter formulation is typical of the structure in which eigenvalue problems often arise in the stability of distributed systems.

8.2 General Features

Suppose that the eigenvalue problem has the general form

$$\mathscr{L}_1[x(z)] - \lambda \mathscr{L}_2[x(z)] = 0 \qquad (8\text{-}5)$$

with homogeneous boundary conditions at $z = 0$ and 1 (i.e., linear combinations of x and its derivatives equal zero at the boundary). $\mathscr{L}_1$ and $\mathscr{L}_2$ are linear differential operators. In Eq. (8-1), $\mathscr{L}_1 = d^4/dx^4$, $\mathscr{L}_2 = 1$. We assume that the eigenfunction can be approximated by

$$x(z) \simeq \sum_{k=1}^{N} C_k \phi_k(z) \qquad (8\text{-}6)$$

The $\phi_k(z)$ are linearly independent functions, each of which satisfies the homogeneous boundary conditions on $x(z)$. Upon substitution into the left side of Eq. (8-5) we get

$$\sum_{k=1}^{N} C_k \mathscr{L}_1[\phi_k(z)] - \lambda \sum_{k=1}^{N} C_k \mathscr{L}_2[\phi_k(z)] = \mathscr{R}(\{C_k\}, z) \neq 0 \qquad (8\text{-}7)$$

The function $\mathscr{R}(\{C_k\}, z)$, which depends explicitly on the coefficients C_k and position z, is the *residual* which results because Eq. (8-6) is not a solution of Eq. (8-5). We can write $\mathscr{L}_i[\sum_k C_k \phi_k] = \sum_k C_k \mathscr{L}_i [\phi_k]$ because of the linearity of the operators $\mathscr{L}_1$ and $\mathscr{L}_2$.

Thus far there is nothing unique about what we have done. The essence of the method is now to establish equations for the coefficients by requiring that the residual be orthogonal to each of the approximating functions,

$$\int_0^1 \mathscr{R}(\{C_k\}, z)\phi_m(z)\, dz = 0, \qquad m = 1, 2, \dots, N \qquad (8\text{-}8)$$

or

$$\sum_{k=1}^{N} \left\{ \int_0^1 \phi_m(z)\mathscr{L}_1[\phi_k(z)]\, dz - \lambda \int_0^1 \phi_m(z)\mathscr{L}_2[\phi_k(z)]\, dz \right\} C_k = 0,$$
$$m = 1, 2, \dots, N \quad (8\text{-}9)$$

Equation (8-9) is a set of N linear, homogeneous algebraic equations for the C_k. A nonzero solution exists if and only if the determinant of coefficients vanishes,

$$|A_{mk} - \lambda B_{mk}| = 0 \qquad (8\text{-}10)$$

$$A_{mk} = \int_0^1 \phi_m(z)\mathscr{L}_1[\phi_k(z)]\, dz$$

$$B_{mk} = \int_0^1 \phi_m(z)\mathscr{L}_2[\phi_k(z)]\, dz$$

Equation (8-10) is an Nth order polynomial equation. Thus, an N-term expansion produces the first N from among the infinite number of eigenvalues of the system (8-5).

The orthogonalization, Eq. (8-8), seems to be a quite arbitrary way of producing a set of equations for the coefficients. We shall show in Sec. 8.6 how it follows in some cases from the calculus of variations. It can be rationalized here by supposing that we let $N \to \infty$ and take the $\phi_k(z)$ to be a complete set of functions, such as $\cos(k - 1)\pi z/2$. The only function which is orthogonal to every member of a complete set is the function which is identically zero. Thus, by this particular choice we shall obtain an exact solution with a zero residual.

8.3 Example

We shall now use Eq. (8-1) as a specific example. The first problem is to choose a set of functions which satisfy all the boundary conditions. If we select polynomials with only even powers, then some preliminary manipulation shows that a set of functions satisfying all four boundary conditions is

$$\phi_k(z) = z^{2k-2}[2 - z^{2k-2}][1 - \tfrac{6}{5}z^2 + \tfrac{1}{3}z^4] \qquad (8\text{-}11)$$

Following the procedure outlined in the previous section we obtain the equations

$$\sum_{k=1}^{N} \left\{ \left[\int_0^1 z^{2m-2}[2-z^{2m-2}] \left[1 - \frac{6}{5}z^2 + \frac{1}{5}z^4 \right] \frac{d^4}{dz^4} z^{2k-2}[2-z^{2k-2}] \right. \right.$$

$$\left. \left[1 - \frac{6}{5}z^2 + \frac{1}{5}z^4 \right] dz - \lambda \int_0^1 z^{2m+2k-4}[2-z^{2m-2}] \left[1 - \frac{6}{5}z^2 + \frac{1}{5}z^4 \right]^2 dz \right\} C_k$$

$$= 0, \qquad m = 1, 2, \ldots, N \quad (8\text{-}12)$$

The simplest case is $N = 1$. Then Eq. (8-12) becomes

$$[3.07 - 0.503\lambda]C_1 = 0$$

which has a nonzero solution for

$$\lambda = 6.10$$

This is a very close approximation to the exact first eigenvalue, $\lambda = 6.09$ (to three significant figures). When $N = 2$ we have

$$[3.07 - 0.503\lambda]C_1 + [0.926 - 0.112\lambda]C_2 = 0$$
$$[0.926 - 0.112\lambda]C_1 + [14.03 - 0.0521\lambda]C_2 = 0$$

A nontrivial solution requires that the determinant of coefficients vanish,

$$[3.07 - 0.503\lambda][14.03 - 0.0521\lambda] - [0.926 - 0.112\lambda]^2 = 0$$

which has roots

$$\lambda = 6.09, \quad 507$$

The exact value of the second eigenvalue to three significant figures is $\lambda = 493$. By substituting back into the linear equations we can obtain the ratio C_2/C_1 for each eigenfunction:

$$x_1(z) = C_{11}[1 - \tfrac{6}{5}z^2 + \tfrac{1}{5}z^4]\{1 - 0.16z^2[2 - z^2] + \cdots\}$$
$$x_2(z) = C_{21}[1 - \tfrac{6}{5}z^2 + \tfrac{1}{5}z^4]\{1 - 4.5z^2[2 - z^2] + \cdots\}$$

C_{11} and C_{21} are arbitrary coefficients, since the eigenfunctions have not been normalized.

A word of caution is in order here. The excellent agreement between the approximate and exact eigenvalues with so few terms is a consequence of the simple structure of the equation and the smooth eigenfunctions. In some of the problems which we shall consider later 1 or 2 terms in the expansion will suffice, while in others as many as 12 to 18 terms will be required to obtain good convergence for the first eigenvalue.

8.4 Coupled Equations

It is often difficult to obtain reasonable functions satisfying all the boundary conditions in a high-order system, and the manipulations involving the higher derivatives can be quite tedious. It is sometimes easier to deal directly with the lower-order coupled system. Consider the example problem in the form of Eq. (8-4). We can write

$$x(z) = \sum_{n=1}^{N} C_n \phi_n(z)$$

$$y(z) = \sum_{m=1}^{M} K_m \psi_m(z)$$

N and M need not be equal. Then, substituting into Eqs. (8-4),

$$\mathcal{R}_1(\{C_n\}, \{K_m\}, z) = \sum_{n=1}^{N} C_n \phi_n''(z) - \sum_{m=1}^{M} K_m \psi_m(z)$$

$$\mathcal{R}_2(\{C_n\}, \{K_m\}, z) = \sum_{m=1}^{M} K_m \psi_m''(z) - \lambda \sum_{n=1}^{N} C_n \phi_n(z)$$

The coefficients are found by taking $\mathcal{R}_1$ orthogonal to the $\phi_n(z)$ and $\mathcal{R}_2$ orthogonal to the $\psi_m(z)$:

$$\int_0^1 \mathcal{R}_1(\{C_n\}, \{K_m\}, z)\phi_p(z)\, dz = 0, \qquad p = 1, 2, \ldots, N$$

$$\int_0^1 \mathcal{R}_2(\{C_n\}, \{K_m\}, z)\psi_q(z)\, dz = 0, \qquad q = 1, 2, \ldots, M$$

For illustrative purposes we shall take as approximating functions

$$\phi_n(z) = z^{2n-2}[1 - z^2]$$
$$\psi_m(z) = z^{2m-2}[1 - z^4]$$

The choice of z^2 and z^4 in the two sets of functions is simply to emphasize the essential arbitrariness of the approximating functions. In fact, a better choice for this problem is to take ϕ_n as given and $\psi_n = \phi_n$. Taking $N = M = 1$ and carrying out the indicated operations we get the equations

$$105C_1 + 64K_1 = 0$$
$$64\lambda C_1 + 240K_1 = 0$$

The determinant must vanish, giving a value of

$$\lambda = 6.15$$

Note that in this approach a two-by-two determinant must be evaluated to find only the first eigenvalue. By contrast, in the preceding section we obtained two eigenvalues from a two-by-two determinant. When a large number of terms is involved this could be an important consideration. The equations will often contain complex numbers, and evaluating the eigenvalues of large complex matrices is a slow operation, even on the most modern digital computers.

8.5 Partial Integration

There is an alternative approach to solving the coupled system which reduces the size of the determinant. Here we choose a set of approximating functions for one variable, solve the differential equation for the second exactly, and then orthogonalize the remaining residual. For example, in Eqs. (8-4) we take

$$x(z) = \sum_{n=1}^{N} C_n z^{2n-2}[1 - z^2]$$

Then the equation for $y(z)$ is

$$y'' = \lambda \sum_{n=1}^{N} C_n z^{2n-2}[1 - z^2]$$

which has the solution

$$y(z) = -\lambda \sum_{n=1}^{N} C_n \left\{ \frac{8n+2}{[2n-1][2n][2n+1][2n+2]} - \frac{z^{2n}}{2n[2n-1]} \right.$$
$$\left. + \frac{z^{2n+2}}{[2n+1][2n+2]} \right\}$$

The residual for the x equation is then

$$\mathcal{R}_1(\{C_n\}, z) = \sum_{n=1}^{N} \left[\frac{d^2}{dz^2} z^{2n-2}[1-z^2] - \lambda \left\{ \frac{8n+2}{[2n-1][2n][2n+1][2n+2]} \right. \right.$$
$$\left. \left. - \frac{z^{2n}}{2n[2n-1]} + \frac{z^{2n+2}}{[2n+1][2n+2]} \right\} \right]$$

and the coefficients are evaluated by the orthogonalization

$$\int_0^1 \mathcal{R}_1(\{C_n\}, z) z^{2m-2}[1-z^2] \, dz = 0, \qquad m = 1, 2, \ldots, N$$

For $N = 1$ we obtain the single equation

$$\{-\tfrac{4}{3} + \tfrac{68}{105}\lambda\}C_1 = 0$$

$$\lambda = 6.18$$

As many eigenvalues are obtained as there are terms in the expansion, but an intermediate differential equation must be solved. This approach is used in Chapter 10 to solve the problem of stability of rotational Couette flow.

8.6 Calculus of Variations

The most elementary problem in the calculus of variations is that of finding the function $x(z)$ with $x(0) = x(1) = 0$ which minimizes the integral

$$\mathcal{E} = \int_0^1 \mathcal{F}(x, x', z) \, dz \tag{8-13}$$

It can be shown that the minimizing function is a solution of the *Euler differential equation*

$$\frac{\partial \mathcal{F}}{\partial x} - \frac{d}{dz} \frac{\partial \mathcal{F}}{\partial x'} = 0 \tag{8-14}$$

$$x(0) = x(1) = 0$$

One way of carrying out the minimization in Eq. (8-13) directly is to seek an approximate solution with N unspecified coefficients

$$x(z) = \sum_{k=1}^{N} C_k \phi_k(z), \qquad \phi_k(0) = \phi_k(1) = 0$$

Then Eq. (8-13) becomes

$$\mathscr{E} = \int_0^1 \mathscr{F}\left(\sum_{k=1}^N C_k \phi_k, \sum_{k=1}^N C_k \phi_k', z\right) dz \qquad (8\text{-}15)$$

Minimization is now a problem in differential calculus, and the minimum is found by setting derivatives with respect to each C_m to zero. This is the *Rayleigh-Ritz method*,

$$\frac{\partial \mathscr{E}}{\partial C_m} = \int_0^1 \left\{ \frac{\partial \mathscr{F}}{\partial x} \phi_m + \frac{\partial \mathscr{F}}{\partial x'} \phi_m' \right\} dz = 0$$

The second term is integrated by parts, and we obtain

$$\frac{\partial \mathscr{E}}{\partial C_m} = \int_0^1 \left\{ \frac{\partial \mathscr{F}}{\partial x} - \frac{d}{dz}\frac{\partial \mathscr{F}}{\partial x'} \right\} \phi_m(z) \, dz = 0 \qquad (8\text{-}16)$$

The quantity in braces is the residual of the Euler equation (8-14). Thus, Eq. (8-16) represents the orthogonality condition for the residual which we have used in Galerkin's method. Whenever the differential equation can be obtained from a minimization or maximization problem (a *variational problem*) Galerkin's method is equivalent to the Rayleigh-Ritz method. This is the basis of the orthogonalization procedure introduced in Eq. (8-8) to find the coefficients in the approximate solution.

8.7 Collocation

Galerkin's method is just one of a group of procedures known as *methods of weighted residuals*, in which equations for the N coefficients are obtained by carrying out N linear operations on the residual. The simplest of these methods, which has attracted considerable interest in recent years, is the method of collocation. Here, the residual is simply set to zero at N collocation points $z_{c1}, z_{c2}, \ldots, z_{cN}$. We shall illustrate with Eq. (8-1) and $N = 1$.

For an approximating function we continue to use Eq. (8-11), which, for $N = 1$, is

$$\phi(z) = 1 - \tfrac{6}{5}z^2 + \tfrac{1}{5}z^4$$

The residual is then

$$\mathscr{R}(C, z) = C[\tfrac{24}{5} - \lambda(1 - \tfrac{6}{5}z^2 + \tfrac{1}{5}z^4)]$$

Setting the residual to zero at the collocation point z_c gives an equation for the eigenvalue,

$$\lambda = \frac{24}{5 - 6z_c^2 + z_c^4}$$

λ is shown as a function of z_c in Fig. 8.1. The dashed horizontal line is the exact first eigenvalue, $\lambda = 6.09$, which is indistinguishable on the figure from the result obtained using Galerkin's method with $N = 1$. It is evident that

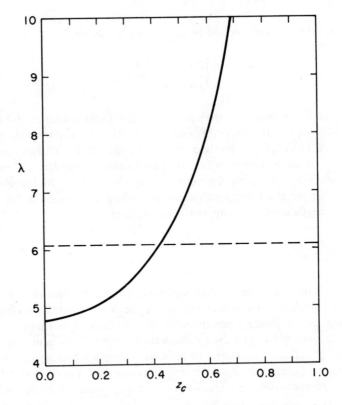

Figure 8.1 Eigenvalue estimate as a function of the collocation point.

considerable care must be taken in the choice of z_c in order to ensure a reasonable estimate of λ. The choice of collocation points is closely related to the approximating functions used. This is discussed in some detail in the literature on the subject.

BIBLIOGRAPHICAL NOTES

The most comprehensive treatment of Galerkin's method and other methods of weighted residuals is in

Finlayson, B. A., *The Method of Weighted Residuals and Variational Principles*, Academic Press, New York, 1972.

Finlayson discusses applications to a variety of problems, including some treated in this text, as well as the convergence of the approximating series. There is an extensive bibliography in his earlier paper,

Finlayson, B. A., and L. E. Scriven, *Appl. Mech. Rev.*, *19*, 735 (1966).

Other texts dealing with these and related methods are

Ames, W. F., *Nonlinear Partial Differential Equations in Engineering*, Academic Press, New York, 1965.

Collatz, L., *The Numerical Treatment of Differential Equations*, Springer, Berlin, 1960.

Denn, M. M., *Optimization by Variational Methods*, McGraw-Hill, New York, 1969.

Kantorovich, L. V., and Krylov, V. I., *Approximate Methods of Higher Analysis*, Wiley-Interscience, New York, 1958.

Schechter, R. S., *The Variational Method in Engineering*, McGraw-Hill, New York, 1967.

Linear Stability: Catalytic Reaction

9

9.1 Introduction

The stability of a catalytic reaction contains many of the general features important in the study of stability of a distributed parameter system to infinitesimal perturbations, so it is a helpful introductory example. The approach which we use is identical to that employed in Chapter 4 for lumped systems, where we searched for conditions under which solutions to the linearized perturbation equations grow in time. A rigorous general proof that the linear equations are truly a valid approximation to the behavior of the nonlinear distributed system does not exist, in contrast to the lumped system. On the other hand, contrary behavior has never been observed in practical applications and would represent pathological behavior not to be expected in real physical situations. Thus, we shall use linearization without hesitation.

We shall restrict attention to the case of an irreversible first-order reaction in a slab catalyst. The steady-state behavior was considered in Chapter 7. The dimensionless mass and energy balance equations are, respectively,

$$\frac{\partial x}{\partial t} = \frac{\partial^2 x}{\partial z^2} - \alpha x e^{-\gamma/y} \tag{9-1}$$

$$\mathscr{L}\frac{\partial y}{\partial t} = \frac{\partial^2 y}{\partial z^2} + \alpha\beta x e^{-\gamma/y} \tag{9-2}$$

$$x = y = 1 \qquad \text{at } z = 1 \tag{9-3a}$$

$$\frac{\partial x}{\partial z} = \frac{\partial y}{\partial z} = 0 \qquad \text{at } z = 0 \tag{9-3b}$$

The steady state satisfies the relation

$$\beta x_s(z) + y_s(z) = 1 + \beta \tag{9-4}$$

9.2 Adiabatic Perturbation, $\mathscr{L} = 1$

There are some very special features associated with the case of unit Lewis number, $\mathscr{L} = 1$, which is within the practical range of that parameter. If we multiply Eq. (9-1) by β and add to Eq. (9-2), we obtain

$$\frac{\partial w}{\partial t} = \frac{\partial^2 w}{\partial z^2} \tag{9-5}$$

$$w = \beta x + y \tag{9-6}$$

with boundary conditions

$$w = 1 + \beta \qquad \text{at } z = 1 \tag{9-7a}$$

$$\frac{\partial w}{\partial z} = 0 \qquad \text{at } z = 0 \tag{9-7b}$$

w is sometimes called the *residual enthalpy*. Equation (9-2) can be written as

$$\frac{\partial y}{\partial t} = \frac{\partial^2 y}{\partial z^2} + \alpha[w - y]e^{-\gamma/y} \tag{9-8}$$

and the pair of equations (9-5) and (9-8) is equivalent to the original set.

A remarkable simplification occurs if the system is initially perturbed from the steady state such that the relationship

$$\beta x + y = 1 + \beta \qquad \text{at } t = 0 \tag{9-9}$$

is preserved. In that case the solution to Eq. (9-5) with boundary conditions (9-7) is a constant, $w = 1 + \beta$, and the system reduces to a single equation,

$$\frac{\partial y}{\partial t} = \frac{\partial^2 y}{\partial z^2} + \alpha F(y) \qquad (9\text{-}10)$$

where $F(y)$ is defined by Eq. (7-7),

$$F(y) = [\beta + 1 - y]e^{-\gamma/y} \qquad (9\text{-}11)$$

We now linearize by defining

$$\eta = y - y_s \qquad (9\text{-}12)$$

in which case Eq. (9-10) becomes

$$\frac{\partial \eta}{\partial t} = y_s'' + \frac{\partial^2 \eta}{\partial z^2} + \alpha F(y_s + \eta)$$

$$= y_s'' + \frac{\partial^2 \eta}{\partial z^2} + \alpha F(y_s) + \alpha F'(y_s)\eta + \cdots$$

The steady-state terms sum to zero, and we have the linear perturbation equation

$$\frac{\partial \eta}{\partial t} = \frac{\partial^2 \eta}{\partial z^2} + \alpha F'(y_s)\eta \qquad (9\text{-}13)$$

Since $y = y_s = 1$ at $z = 1$, we obtain the boundary conditions

$$\frac{\partial \eta}{\partial z} = 0 \qquad \text{at } z = 0, \qquad \eta = 0 \qquad \text{at } z = 1 \qquad (9\text{-}14)$$

Equation (9-13) is a linear homogeneous partial differential equation with linear homogeneous boundary conditions. It can be solved by a variety of methods, including Laplace transform in time and (finite) Fourier transform in the spatial variable. The latter is the most efficient. It is likely, however, that readers of this book are most familiar with the method of *separation of variables*, where we seek a solution of the form

$$\eta = \theta(t)\phi(z) \qquad (9\text{-}15)$$

$\phi(z)$ must satisfy the boundary conditions

$$\phi'(0) = \phi(1) = 0 \tag{9-16}$$

The solution to Eq. (9-13) is a linear combination of all solutions of the form (9-15), where the coefficients in the linear combination are chosen to satisfy the initial conditions.

Substituting Eq. (9-15) into (9-13) gives

$$\dot{\theta}\phi = \theta\phi'' + \alpha F'(y_s)\theta\phi$$

where a dot denotes differentiation with respect to t. Equivalently,

$$\frac{\dot{\theta}}{\theta} = \frac{\phi'' + \alpha F'(y_s)\phi}{\phi} = \lambda \tag{9-17}$$

The first ratio in Eq. (9-17) is a function only of t, and the second only of z. If they are equal to each other, they must both equal a constant, λ. Thus,

$$\dot{\theta} - \lambda\theta = 0 \tag{9-18}$$

$$\phi'' + [-\lambda + \alpha F'(y_s)]\phi = 0 \tag{9-19}$$

λ is an eigenvalue of Eq. (9-19). The solution to Eq. (9-18) is

$$\theta = e^{\lambda t} \tag{9-20}$$

(Any coefficient can be included in ϕ.) Thus, the system will be unstable to infinitesimal perturbations if Eq. (9-19), with boundary conditions (9-13), has *any* eigenvalue with a positive real part. Instability occurs with even a single positive eigenvalue because we take a linear combination of all eigensolutions to form the general solution to Eq. (9-13).

The properties of Eq. (9-19) are established in Appendix D. The eigenvalues are all real, so an oscillatory response is impossible. Furthermore, the largest eigenvalue satisfies the inequality

$$\lambda \leq \alpha \max F'(y_s) - \frac{\pi^2}{4}$$

Thus, the eigenvalues will all be negative, and perturbations will damp out if the steady-state solution is such that

$$\alpha \max F'(y_s) \leq \frac{\pi^2}{4} \tag{9-21}$$

Note that the maximum here is only over values of y_s along the steady state, so the stability criterion can be satisfied even when the uniqueness criterion (7-19) is not.

If we wish the actual eigenvalues for a given steady state in order to estimate the growth or decay of the transient, we must use some approximating scheme, since Eq. (9-19) does not admit an analytical solution in terms of simple functions. The choice of approximating functions for Galerkin's method is suggested by the form of Eq. (9-19), which would have cosine solutions for $F'(y_s) = $ constant. Thus, we take

$$\phi \simeq \sum_{n=1}^{N} C_n \cos \frac{2n-1}{2} \pi z \qquad (9\text{-}22)$$

$$\phi'' \simeq - \sum_{n=1}^{N} \left[\frac{2n-1}{2} \right]^2 \pi^2 C_n \cos \frac{2n-1}{2} \pi z$$

Following Sec. 8.2 we obtain the residual

$$\mathcal{R}(\{C_n\}, z) = - \sum_{n=1}^{N} \left[\frac{2n-1}{2} \right]^2 \pi^2 C_n \cos \frac{2n-1}{2} \pi z - \lambda \sum_{n=1}^{N} C_n \cos \frac{2n-1}{2} \pi z$$

$$+ \alpha F'(y_s) \sum_{n=1}^{N} C_n \cos \frac{2n-1}{2} \pi z \qquad (9\text{-}23)$$

The orthogonality condition,

$$\int_0^1 \mathcal{R}(\{C_n\}, z) \cos \frac{2m-1}{2} \pi z \, dz = 0, \qquad m = 1, 2, \dots, N$$

then leads to the equations

$$\sum_{n=1}^{N} A_{mn} C_n - \lambda C_m = 0, \qquad m = 1, 2, \dots, N \qquad (9\text{-}24)$$

$$A_{mn} = 2\alpha \int_0^1 F'(y_s(z)) \cos \frac{2m-1}{2} \pi z \cos \frac{2n-1}{2} \pi z \, dz - \left[\frac{2m-1}{2} \right]^2 \pi^2 \delta_{mn}$$

$$\delta_{mn} = \begin{cases} 1, & m = n \\ 0, & m \neq n \end{cases}$$

The eigenvalues λ are therefore eigenvalues of the matrix A_{mn}.

Convergence is quite rapid with respect to N. For the parameters given in Sec. 7.4 and the steady-state profiles in Fig. 7.2 the following results were obtained by J. Wei:

	Low Temperature	Middle Temperature	High Temperature
$N = 1$, $\lambda_1 = -2.08$		$+3.94$	-75.53
$N = 2$, $\lambda_1 = -2.08$		$+4.18$	-28.32
$\lambda_2 = -21.83$		-18.15	-101.38

λ_1 is essentially converged to its final value by $N = 2$ for the first two steady states. At the high-temperature state C. R. McGowin's calculations indicate essential convergence at $N = 3$ to $\lambda_1 = -18.82$, with a value $\lambda_1 = -18.26$ for $N = 10$. The middle steady state is clearly unstable; the other two are stable to infinitesimal disturbances.

There is one serious restriction on the results obtained in this section. We have assumed a special form of infinitesimal perturbation, the adiabatic perturbation. Thus, any state which we establish as unstable is certainly unstable. If we compute a state as stable, however, we cannot be certain that it will remain stable to nonadiabatic infinitesimal disturbances. Hence, inequality (9-21) establishes a bound beyond which instability must occur, but it may not be the best bound possible even for infinitesimal disturbances.

9.3 $\mathscr{L} = 1$

It is possible to prove that the adiabatic perturbation is the worst perturbation for the catalytic reaction with $\mathscr{L} = 1$, in that it gives the stability condition corresponding to the case of an arbitrary infinitesimal disturbance. This is a very important point. We often make explicit assumptions about the type of disturbance in order to obtain a solution, but only in some cases can we establish that the result is valid for all disturbances.

We return again to Eqs. (9-5) and (9-8) and define two deviation variables,

$$\eta = y - y_s$$
$$\omega = w - w_s = w - 1 - \beta \qquad (9\text{-}25)$$

Substituting into the two equations, expanding the nonlinearity in Eq. (9-8), and retaining only linear terms leads to the perturbation equations

$$\frac{\partial \eta}{\partial t} = \frac{\partial^2 \eta}{\partial z^2} + \alpha F'(y_s)\eta + \alpha e^{-\gamma/y_s}\omega \qquad (9\text{-}26)$$

$$\frac{\partial \omega}{\partial t} = \frac{\partial^2 \omega}{\partial z^2} \tag{9-27}$$

$$\frac{\partial \eta}{\partial z} = \frac{\partial \omega}{\partial z} = 0 \qquad \text{at } z = 0 \tag{9-28a}$$

$$\eta = \omega = 0 \qquad \text{at } z = 1 \tag{9-28b}$$

If we solve this set of coupled equations by separation of variables, we write

$$\eta = \theta_1(t)\phi(z) \tag{9-29a}$$

$$\omega = \theta_2(t)\psi(z) \tag{9-29b}$$

$$\phi'(0) = \psi'(0) = \phi(1) = \psi(1) = 0 \tag{9-30}$$

Equation (9-26) then becomes

$$\dot{\theta}_1 \phi = \theta_1 \phi'' + \alpha F'(y_s)\theta_1 \phi + \alpha e^{-\gamma/y_s}\theta_2 \psi$$

or

$$\frac{\dot{\theta}_1}{\theta_1} = \frac{\phi'' + \alpha F'(y_s)\phi + \alpha e^{-\gamma/y_s}\psi[\theta_2/\theta_1]}{\phi} = \lambda \tag{9-31}$$

The second ratio is independent of t only when $\theta_2 = \theta_1$. (Any proportionality constant can be taken up in ψ.) Thus, *solution by separation of variables is possible if and only if*

$$\theta_1 = \theta_2 = e^{\lambda t} \tag{9-32}$$

This is a general result, and we shall use it henceforth without repeating the steps in the derivation. In a set of coupled linear homogeneous partial differential equations with coefficients that are independent of time, the time dependence of each dependent variable in a separation of variables solution is of the form exp (λt).

Making use of Eq. (9-32) the linear equations (9-26) and (9-27) reduce to two ordinary differential equations,

$$\phi'' + [-\lambda + \alpha F'(y_s)]\phi + \alpha e^{-\gamma/y_s}\psi = 0 \tag{9-33}$$

$$\psi'' - \lambda\psi = 0 \tag{9-34}$$

Two sets of eigensolutions are possible. The first, in which $\psi = 0$, $\phi \neq 0$, corresponds to the adiabatic perturbation which we have already studied.

If $\psi \neq 0$, then we can solve Eq. (9-34) with boundary conditions (9-30) analytically:

$$\psi(z) = \mathscr{A} \cosh \lambda^{1/2} z = \mathscr{A} \cos i\lambda^{1/2} z \qquad (9\text{-}35)$$

The condition $\psi(1) = 0$ requires that λ take on values

$$i\lambda^{1/2} = \frac{2n-1}{2} \pi$$

$$\lambda = -\left[\frac{2n-1}{2}\right]^2 \pi^2, \qquad n = 1, 2, \ldots \qquad (9\text{-}36)$$

For these values of λ and ψ a nonzero solution of Eq. (9-33) can then be obtained. All eigenvalues defined by Eq. (9-36) are negative and correspond to decaying exponential modes. Thus, the only disturbance modes which can grow are those corresponding to the eigenvalues of Eq. (9-33) when $\psi = 0$. The adiabatic perturbation is therefore the only one which need be considered for this problem.

9.4 $\mathscr{L} \neq 1$

When the Lewis number is not equal to unity the stability problem cannot be treated analytically in the same simple fashion, but results for specific parameters can be obtained without serious difficulty. It is again convenient to use the residual enthalpy defined by Eq. (9-6),

$$w = \beta x + y$$

By multiplying Eq. (9-1) by β and adding to Eq. (9-2) we obtain the equivalent pair of equations

$$\frac{\partial w}{\partial t} + [\mathscr{L} - 1] \frac{\partial y}{\partial t} = \frac{\partial^2 w}{\partial z^2} \qquad (9\text{-}37)$$

$$\mathscr{L} \frac{\partial y}{\partial t} = \frac{\partial^2 y}{\partial z^2} + \alpha[w - y]e^{-\gamma/y} \qquad (9\text{-}38)$$

Then, using the definitions (9-25) for the deviation variables,

$$\eta = y - y_s, \qquad \omega = w - 1 - \beta$$

we obtain the equivalent linearized system by neglecting quadratic and higher terms,

$$\frac{\partial \omega}{\partial t} + [\mathscr{L} - 1]\frac{\partial \eta}{\partial t} = \frac{\partial^2 \omega}{\partial z^2} \qquad (9\text{-}39)$$

$$\mathscr{L}\frac{\partial \eta}{\partial t} = \frac{\partial^2 \eta}{\partial z^2} + \alpha F'(y_s)\eta + \alpha e^{-\gamma/y_s}\omega \qquad (9\text{-}40)$$

We seek a solution by separation of variables in the form

$$\omega = e^{\lambda t}\psi(z) \qquad \eta = e^{\lambda t}\phi(z) \qquad (9\text{-}41)$$

in which case Eqs. (9-39) and (9-40) become

$$\psi'' - \lambda\psi - \lambda[\mathscr{L} - 1]\phi = 0 \qquad (9\text{-}42)$$

$$\phi'' + [-\mathscr{L}\lambda + \alpha F'(y_s)]\phi + \alpha e^{-\gamma/y_s}\psi = 0 \qquad (9\text{-}43)$$

$$\psi'(0) = \phi'(0) = \psi(1) = \phi(1) = 0 \qquad (9\text{-}44)$$

The structure is revealed a bit more fully by differentiating Eq. (9-42) twice and eliminating ϕ by use of Eq. (9-43) to obtain

$$\psi^{IV} + \{-[1 + \mathscr{L}]\lambda + \alpha F'(y_s)\}\psi'' + \lambda\{\mathscr{L}\lambda - \alpha F'(y_s) + \alpha[\mathscr{L} - 1]e^{-\gamma/y_s}\}\psi = 0$$
$$(9\text{-}45a)$$

or, equivalently,

$$\left\{\frac{d^2}{dz^2} - \mathscr{L}\lambda + \alpha F'(y_s)\right\}\left\{\frac{d^2}{dz^2} - \lambda\right\}\psi + \lambda[\mathscr{L} - 1]\alpha e^{-\gamma/y_s}\psi = 0 \qquad (9\text{-}45b)$$

The boundary conditions follow from Eq. (9-44) as

$$\psi'(0) = \psi'''(0) = \psi(1) = \psi''(1) = 0 \qquad (9\text{-}46)$$

When $\mathscr{L} \to 1$ we easily see that the two sets of eigenvalues found in the preceding section arise from the product of the two differential operators. Equation (9-45) is regular in the parameter $\mathscr{L} - 1$, so solutions in the neighborhood of Lewis number 1 can be obtained by regular perturbation methods, though we shall not pursue that idea here. Equation (9-45) is not self-adjoint (Appendix E), and we cannot prove that the eigenvalues are real for $\mathscr{L} \neq 1$. In fact, complex eigenvalues are found for some values of $\mathscr{L}$.

The eigenvalues can be estimated using Galerkin's method. We seek a solution in the form

$$\psi(z) \simeq \sum_{n=1}^{N} C_n \cos \frac{2n-1}{2} \pi z \tag{9-47}$$

$$\psi'' \simeq -\sum_{n=1}^{N} C_n \left[\frac{2n-1}{2}\right]^2 \pi^2 \cos \frac{2n-1}{2} \pi z$$

$$\psi^{\text{IV}} \simeq \sum_{n=1}^{N} C_n \left[\frac{2n-1}{2}\right]^4 \pi^4 \cos \frac{2n-1}{2} \pi z$$

Substituting into Eq. (9-45) gives the residual

$$\mathscr{R}(\{C_n\}, z) = \sum_{n=1}^{N} C_n \left\{ \left[\frac{2n-1}{2}\right]^4 \pi^4 - \{-[1 + \mathscr{L}]\lambda + \alpha F'(y_s)\}\left[\frac{2n-1}{2}\right]^2 \pi^2 \right.$$

$$\left. + \lambda\{\mathscr{L}\lambda - \alpha F'(y_s) + \alpha[\mathscr{L} - 1]e^{-\gamma/y_s}\} \right\} \cos \frac{2n-1}{2} \pi z \tag{9-48}$$

and from the orthogonality condition,

$$\int_0^1 \mathscr{R}(\{C_n\}, z) \cos \frac{2m-1}{2} \pi z \, dz = 0, \qquad m = 1, 2, \ldots, N$$

we obtain the homogeneous algebraic equations

$$\sum_{n=1}^{N} B_{mn} C_n = 0, \qquad m = 1, 2, \ldots, N \tag{9-49}$$

$$B_{mn} = -\alpha\left\{ \left[\frac{2n-1}{2}\right]^2 \pi^2 + \lambda \right\} \int_0^1 F'(y_s) \cos \frac{2m-1}{2} \pi z \cos \frac{2n-1}{2} \pi z \, dz$$

$$+ \alpha[\mathscr{L} - 1] \int_0^1 e^{-\gamma/y_s} \cos \frac{2m-1}{2} \pi z \cos \frac{2n-1}{2} \pi z \, dz$$

$$+ \left\{ \left[\frac{2n-1}{2}\right]^4 \frac{\pi^4}{2} + \lambda[1 + \mathscr{L}]\left[\frac{2n-1}{2}\right]^2 \frac{\pi^2}{2} + \frac{\mathscr{L}\lambda^2}{2} \right\} \delta_{mn}$$

$$\delta_{mn} = \begin{cases} 0, & m \neq n \\ 1, & m = n \end{cases}$$

The eigenvalues, λ, are found from the condition that the determinant of B_{mn} vanish. For example, if we take $N = 1$, we get the equation

$$\frac{\lambda^2 \mathscr{L}}{2} + \lambda \left\{ [1 + \mathscr{L}] \frac{\pi^2}{8} - \alpha \int_0^1 F'(y_s) \cos^2 \frac{\pi z}{2} \, dz \right.$$

$$\left. + \alpha[\mathscr{L} - 1] \int_0^1 e^{-\gamma/y_s} \cos^2 \frac{\pi z}{2} \, dz \right\} + \frac{\pi^4}{32} - \frac{\alpha \pi^2}{4} \int_0^1 F'(y_s) \cos^2 \frac{\pi z}{2} \, dz = 0$$

$$(9.50)$$

When $\mathscr{L} = 1$ this gives the pair of eigenvalues $\lambda = -\pi^2/4$, $\lambda = -\pi^2/4 + 2\alpha \int_0^1 F'(y_s) \cos^2 (\pi z/2) \, dz$, corresponding to the first eigenvalue in each of the sequences defined by Eqs. (9-36) and (9-24), respectively.

The first estimate of the stability conditions is obtained by noting (Sec. 4.1) that the roots of Eq. (9-50) will have negative real parts if and only if the coefficient of λ and the constant term are both positive. The integrals

$$2 \int_0^1 F'(y_s) \cos^2 \frac{\pi z}{2} \, dz \quad \text{and} \quad 2 \int_0^1 e^{-\gamma/y_s} \cos^2 \frac{\pi z}{2} \, dz$$

are cosine square-weighted averages, which we denote by $\langle F'(y_s) \rangle$ and $\langle e^{-\gamma/y_s} \rangle$, respectively. The stability conditions are then

$$\alpha \langle F'(y_s) \rangle < \frac{\pi^2}{4} \tag{9-51}$$

$$\alpha \langle F'(y_s) \rangle < \frac{\pi^2}{4} + \mathscr{L} \frac{\pi^2}{4} + \alpha[\mathscr{L} - 1] \langle e^{-\gamma/y_s} \rangle \tag{9-52}$$

The first condition, Eq. (9-51), is an approximation to the exact condition (9-21), where the maximum of $F'(y_s)$ replaces the average. This is independent of Lewis number. We therefore conclude that a steady-state profile which violates Eq. (9-21) and is unstable for $\mathscr{L} = 1$ is unstable for any Lewis number. For $\mathscr{L} \geq 1$, Eq. (9-51) implies Eq. (9-52). Thus, a profile which is stable for $\mathscr{L} = 1$ is also stable for $\mathscr{L} > 1$. On the other hand, for $\mathscr{L}$ sufficiently less than unity it is possible for Eq. (9-51) to be satisfied and yet for Eq. (9-52) to be violated. The critical value of $\mathscr{L}$, denoted $\mathscr{L}_c$, occurs for equality in Eq. (9-52),

$$\mathscr{L}_c = \frac{\alpha \langle F'(y_s) \rangle - (\pi^2/4) + \alpha \langle e^{-\gamma/y_s} \rangle}{(\pi^2/4) + \alpha \langle e^{-\gamma/y_s} \rangle} < 1 \tag{9-53}$$

When a positive value of $\mathscr{L}_c$ exists a steady state which satisfies condition (9-51) will be unstable for $\mathscr{L} < \mathscr{L}_c$. Note that a unique steady state which satisfies Eq. (7-19) could still fail to satisfy Eq. (9-52) and be unstable.

The results in the preceding paragraph are only suggestive, for we have already seen that as many as three terms in the Galerkin expansion will be required at the high-temperature steady state, but the structure of the system is clearly revealed. For the steady-state profiles in Fig. 7.2 McGowin found that the dominant eigenvalue corresponding to the high-temperature steady state became complex for $\mathscr{L}$ somewhere between 0.30 and 0.50 and the real part became positive for $\mathscr{L} < \mathscr{L}_c = 0.265$. This is a rather small value of the Lewis number for a real catalyst system, however.

BIBLIOGRAPHICAL NOTES

The subject of this chapter, with extensions and related problems, is covered in great detail with a comprehensive bibliography in Aris's monograph,

> Aris, R., *The Mathematical Theory of Diffusion and Reaction in Permeable Catalysts*, Oxford University Press, Inc., New York, 1974.

See also

> Gavalas, G. R., *Nonlinear Differential Equations of Chemically Reacting Systems*, Springer-Verlag New York, Inc., New York, 1968.
> Perlmutter, D. D., *Stability of Chemical Reactors*, Prentice-Hall, Englewood Cliffs, N.J., 1972.

Perlmutter also considers the similar problem of stability of a plug flow tubular reactor. Related problems in combustion are treated by

> Frank-Kamenetskii, D. A., *Diffusion and Heat Transfer in Chemical Kinetics*, 2nd ed., Plenum, New York, 1969.

The numerical values in the text were computed by

> McGowin, C. R., *The Stability Analysis of Distributed Parameter Chemical Reactors Using Weighted Residual Techniques*, Ph.D. Dissertation, University of Pennsylvania, Philadelphia, 1969.
> Wei, J., *Chem. Eng. Sci.*, *20*, 729 (1965).

Other calculations are in

> Kuo, J. C. W., and N. R. Amundson, *Chem. Eng. Sci.*, *22*, 1185 (1967).
> Lee, J. C. M., and D. Luss, *AIChE J.*, *16*, 620 (1970).

Eigenvalue calculations using the method of collocation are in

> Van den Bosch, B., and L. Padmanabhan, *Chem. Eng. Sci.*, *29*, 805 (1974).

McGowin has also applied the method of collocation extensively, and many of these calculations are summarized in the book by Perlmutter. Further references to applications are tabulated in survey papers in the proceedings of the first and second international symposia on chemical reaction engineering, published, respectively, as

> *Chemical Reaction Engineering, Advances in Chemistry Series No. 109,* American Chemical Society, Washington, D.C., 1972.
> *Chemical Reaction Engineering,* Elsevier, Amsterdam, 1972.

and in

> Denn, M. M., *Ind. Eng. Chem., 61,* no. 2, 46 (1969).
> Denn, M. M., in V. W. Weekman, Jr., ed., *Annual Reviews of Industrial and Engineering Chemistry, 1970,* American Chemical Society, Washington, D.C., 1972.

The problem of linearization in functional equations is dealt with in

> Krasnosel'skii, M. A., *Topological Methods in the Theory of Nonlinear Integral Equations,* Pergamon, Elmsford, N.Y., 1964.

Linear Stability: Rotational Couette Flow   **10**

10.1 Introduction

We now return to the problem of flow transition between long rotating concentric cylinders described in Sec. 1.3. This transition was first successfully treated by Taylor and is often called the *Taylor stability* problem. The basic conservation equations which are valid for any incompressible fluid are the *Cauchy momentum equation*,

$$\rho \frac{\partial \tilde{\mathbf{v}}}{\partial \tilde{t}} + \rho \tilde{\mathbf{v}} \cdot \tilde{\nabla} \tilde{\mathbf{v}} = -\tilde{\nabla} \tilde{p} + \tilde{\nabla} \cdot \tilde{\boldsymbol{\tau}} + \rho \tilde{\mathbf{f}} \qquad (10\text{-}1)$$

and the *continuity equation*,

$$\tilde{\nabla} \cdot \tilde{\mathbf{v}} = 0 \qquad (10\text{-}2)$$

The tilde (~) here denotes a quantity with dimensions, and the letter without a tilde will be used subsequently for the corresponding dimensionless variable. $\mathbf{v}$ is the velocity vector, ρ the density, p the isotropic pressure, $\boldsymbol{\tau}$ the extra-stress tensor, and $\mathbf{f}$ the body force (e.g., gravity). ∇ is the gradient operator.

For an incompressible Newtonian fluid the extra stress is proportional to the symmetric part of the velocity gradient,

$$\tilde{\tau} = \mu[\tilde{\nabla}\tilde{v} + (\tilde{\nabla}\tilde{v})^T] \tag{10-3}$$

μ is the viscosity and $(\ )^T$ denotes transpose. Equation (10-1) then becomes the *Navier-Stokes equation*,

$$\rho\frac{\partial \tilde{v}}{\partial \tilde{t}} + \rho\tilde{v}\cdot\tilde{\nabla}\tilde{v} = -\tilde{\nabla}\cdot\tilde{p} + \mu\tilde{\nabla}^2\tilde{v} + \rho\tilde{f} \tag{10-4}$$

The boundary conditions are that the velocity be bounded and that the fluid adhere to solid walls. At a fluid-fluid interface the velocity and normal force must be continuous through the surface.

Equations (10-3) and (10-4) are made dimensionless by introducing a characteristic velocity, U, and characteristic length scale, H. Then dimensionless variables are defined:

$$\mathbf{v} = \frac{\tilde{v}}{U} \qquad \nabla = H\tilde{\nabla} \qquad p = \frac{\tilde{p}}{\rho U^2} \qquad t = \tilde{t}\frac{U}{H} \qquad \mathbf{f} = \frac{\tilde{f}H}{U^2}$$

The Navier-Stokes and continuity equations are then written as

$$\frac{\partial \mathbf{v}}{\partial t} + \mathbf{v}\cdot\nabla\mathbf{v} = -\nabla p + \frac{1}{\text{Re}}\nabla^2\mathbf{v} + \mathbf{f} \tag{10-5}$$

$$\nabla\cdot\mathbf{v} = 0 \tag{10-6}$$

The *Reynolds number*, Re, is defined as

$$\text{Re} = \frac{HU\rho}{\mu} \tag{10-7}$$

It can be interpreted as the ratio of inertial forces to viscous forces. The limit Re $\rightarrow \infty$ corresponds to an inviscid (*ideal*) fluid. Note that the Navier-Stokes equation contains only a quadratic nonlinearity, the $\mathbf{v}\cdot\nabla\mathbf{v}$ term. This is particularly significant in the application of Liapunov's direct method in Chapter 15.

The flow geometry of interest here is shown in Fig. 10.1. The inner cylinder has radius R and angular velocity Ω_1, and the outer cylinder has

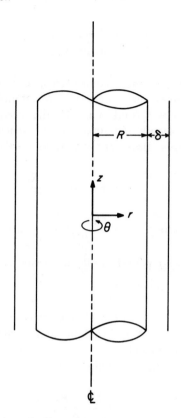

Figure 10.1 Schematic diagram for flow between rotating cylinders.

radius $R + \delta$ and angular velocity Ω_2. In a cylindrical (r, θ, z) coordinate system the Navier-Stokes and continuity equations, (10-5) and (10-6), become

$$\frac{\partial v_r}{\partial t} + v_r \frac{\partial v_r}{\partial r} + \frac{v_\theta}{r} \frac{\partial v_r}{\partial \theta} - \frac{v_\theta^2}{r} + v_z \frac{\partial v_r}{\partial z}$$

$$= -\frac{\partial p}{\partial r} + \frac{1}{Re} \left[\frac{\partial}{\partial r} \frac{1}{r} \frac{\partial}{\partial r} r v_r + \frac{1}{r^2} \frac{\partial^2 v_r}{\partial \theta^2} - \frac{2}{r^2} \frac{\partial v_\theta}{\partial \theta} + \frac{\partial^2 v_r}{\partial z^2} \right] \quad (10\text{-}8a)$$

$$\frac{\partial v_\theta}{\partial t} + v_r \frac{\partial v_\theta}{\partial r} + \frac{v_\theta}{r} \frac{\partial v_\theta}{\partial \theta} + \frac{v_r v_\theta}{r} + v_z \frac{\partial v_\theta}{\partial z}$$

$$= -\frac{1}{r} \frac{\partial p}{\partial \theta} + \frac{1}{Re} \left[\frac{\partial}{\partial r} \frac{1}{r} \frac{\partial}{\partial r} r v_\theta + \frac{1}{r^2} \frac{\partial^2 v_\theta}{\partial \theta^2} + \frac{2}{r^2} \frac{\partial v_r}{\partial \theta} + \frac{\partial^2 v_\theta}{\partial z^2} \right] \quad (10\text{-}8b)$$

$$\frac{\partial v_z}{\partial t} + v_r \frac{\partial v_z}{\partial r} + \frac{v_\theta}{r} \frac{\partial v_z}{\partial \theta} + v_z \frac{\partial v_z}{\partial z} = -\frac{\partial p}{\partial z} + \frac{1}{Re} \left[\frac{1}{r} \frac{\partial}{\partial r} r \frac{\partial v_z}{\partial r} + \frac{1}{r^2} \frac{\partial^2 v_z}{\partial \theta^2} + \frac{\partial^2 v_z}{\partial z^2} \right]$$

$$(10\text{-}8c)$$

$$\frac{1}{r}\frac{\partial}{\partial r}\,rv_r + \frac{1}{r}\frac{\partial v_\theta}{\partial \theta} + \frac{\partial v_z}{\partial z} = 0 \tag{10-8d}$$

v_r, v_θ, and v_z are the r, θ, and z components, respectively, of the velocity vector, $\mathbf{v}$, and body forces, such as gravity, are assumed to be unimportant. The characteristic velocity U is taken as $R\Omega_1$, the speed of the inner cylinder, and the characteristic length H is taken as the spacing, δ. The Reynolds number is Re $= R\Omega_1\delta\rho/\mu$. Boundary conditions are then

$$v_r = v_z = 0, \qquad v_\theta = 1 \qquad \text{at } r = \frac{R}{\delta} \tag{10-9a}$$

$$v_r = v_z = 0, \qquad v_\theta = \frac{[R+\delta]\Omega_2}{R\Omega_1} \qquad \text{at } r = \frac{1+R}{\delta} \tag{10-9b}$$

$$v_r, v_z, v_\theta, p \text{ periodic in } \theta \text{ with period } 2\pi \tag{10-9c}$$

$$v_r, v_z, v_\theta, p \text{ finite as } z \to \pm\infty \tag{10-9d}$$

The term v_θ^2/r in Eq. (10-8a) is the *Coriolis force*, which arises in rotating systems. This is the source of the instability and should be followed carefully in the subsequent development.

It is convenient to take the origin of the radial coordinate at the inner cylinder by defining

$$x = \frac{r-R}{\delta} \tag{10-10}$$

A steady-state solution to Eqs. (10-8) and (10-9) is then readily shown to be

$$v_{rs} = v_{zs} = 0 \tag{10-11a}$$

$$v_{\theta s} \equiv V = \frac{(\Omega_2/\Omega_1) - [R/(R+\delta)]^2}{1 - [R/(R+\delta)]^2}\left[1 + \frac{\delta x}{R}\right] + \frac{1 - (\Omega_2/\Omega_1)}{1 - [R/(R+\delta)]^2}\frac{1}{1+(\delta x/R)} \tag{10-11b}$$

$$p_s = \int_0^x \frac{V^2}{1+(\delta x/R)}\,dx + \text{constant} \tag{10-11c}$$

Note that the solution is independent of the Reynolds number. We shall deal in this text only with the case in which δ/R is small. To first order in δ/R the steady-state velocity can be written as

$$v_{\theta s} \equiv V = \left\{1 - \left[1 - \frac{\Omega_2}{\Omega_1}\right]x\right\}\left\{1 + \frac{\delta x}{R}\right\} + o\!\left(\frac{\delta}{R}\right) \tag{10-12}$$

The steady-state velocity field has circular streamlines which lie in the plane normal to the axis of rotation. This flow pattern is observed at low rotational speeds.

10.2 Perturbation Equations

We shall make two key assumptions in deriving the linear perturbation equations. First, we shall assume that the disturbance motion is axisymmetric, so all $\partial/\partial\theta$ terms are zero. This assumption turns out to be valid experimentally as long as $\Omega_2/\Omega_1 > -0.73$. Second, we shall assume that δ/R is very small. In that case we can write

$$\frac{\partial}{\partial r}\frac{1}{r}\frac{\partial}{\partial r}r[\ ] = \frac{\partial}{\partial x}\frac{1}{1+(Rx/\delta)}\frac{\partial}{\partial x}\left[1+\frac{Rx}{\delta}\right][\ ] \simeq \frac{\partial^2[\ ]}{\partial x^2} \qquad (10\text{-}13a)$$

$$\frac{1}{r}\frac{\partial}{\partial r}r\frac{\partial[\ ]}{\partial r} = \frac{1}{1+(Rx/\delta)}\frac{\partial}{\partial x}\left[1+\frac{Rx}{\delta}\right]\frac{\partial[\ ]}{\partial x} \simeq \frac{\partial^2[\ ]}{\partial x^2} \qquad (10\text{-}13b)$$

$$\frac{1}{r}\frac{\partial}{\partial r}r[\ ] = \frac{1}{1+(Rx/\delta)}\frac{\partial}{\partial x}\left[1+\frac{Rx}{\delta}\right][\ ] \simeq \frac{\partial[\ ]}{\partial x} \qquad (10\text{-}13c)$$

and we use Eq. (10-12) for V in place of the exact equation (10-11b). If we define perturbation variables as

$$u = v_r - v_{rs}, \qquad v = v_\theta - v_{\theta s}$$
$$w = v_z - v_{zs}, \qquad q = p - p_s$$

and substitute Eqs. (10-11a), (10-11c), (10-12), and (10-13) into the Navier-Stokes and continuity equations, (10-8), neglecting quadratic terms we obtain

$$\frac{\partial u}{\partial t} - \frac{2\delta}{R}[1 - Mx]v = -\frac{\partial q}{\partial x} + \frac{1}{Re}\left[\frac{\partial^2 u}{\partial x^2} + \frac{\partial^2 u}{\partial z^2}\right] \qquad (10\text{-}14a)$$

$$\frac{\partial v}{\partial t} - Mu = \frac{1}{Re}\left[\frac{\partial^2 v}{\partial x^2} + \frac{\partial^2 v}{\partial z^2}\right] \qquad (10\text{-}14b)$$

$$\frac{\partial w}{\partial t} = -\frac{\partial q}{\partial z} + \frac{1}{Re}\left[\frac{\partial^2 w}{\partial x^2} + \frac{\partial^2 w}{\partial z^2}\right] \qquad (10\text{-}14c)$$

$$\frac{\partial u}{\partial x} + \frac{\partial w}{\partial z} = 0 \qquad (10\text{-}14d)$$

We have introduced the shorthand

$$M = 1 - \frac{\Omega_2}{\Omega_1} \qquad (10\text{-}15)$$

Boundary conditions derived from Eqs. (10-9a), (10-9b), and (10-9d) are

$$u = v = w = 0 \qquad \text{at } x = 0, 1 \qquad (10\text{-}16a)$$

$$u, v, w, p \text{ finite as } z \to \pm\infty \qquad (10\text{-}16b)$$

The Coriolis term is the term containing δ/R in Eq. (10-14a), and it is retained because of its physical importance despite the small δ/R assumption. It should be emphasized that the small gap assumption is for convenience only, and a solution can be obtained without this assumption.

We shall seek a solution to Eqs. (10-14) and (10-16) by separation of variables, where there are now three independent variables, x, z, and t. Proceeding exactly as in Sec. 9.3 it can be shown that all dependent variables will have a z dependence of the form $\exp{(\beta z)}$. To satisfy the boundedness condition for $z \to \pm\infty$ it is necessary that β be an imaginary number, and we write $\beta = ik$. (Had we used a Fourier transform on z, k would correspond to the Fourier wave number, and we shall use that terminology.) We therefore seek solutions of the form

$$u = e^{\lambda t}e^{ikz}\Phi(x) \qquad (10\text{-}17a)$$

$$v = e^{\lambda t}e^{ikz}\psi(x) \qquad (10\text{-}17b)$$

$$w = e^{\lambda t}e^{ikz}\omega(x) \qquad (10\text{-}17c)$$

$$q = e^{\lambda t}e^{ikz}\eta(x) \qquad (10\text{-}17d)$$

(We have used Φ here in order to reserve ϕ for later use.) Equations (10-14) then become

$$\lambda\Phi - \frac{2\delta}{R}[1 - Mx]\psi = -\eta' + \frac{1}{\text{Re}}[\Phi'' - k^2\Phi] \qquad (10\text{-}18a)$$

$$\lambda\psi - M\Phi = \frac{1}{\text{Re}}[\psi'' - k^2\psi] \qquad (10\text{-}18b)$$

$$\lambda\omega = -ik\eta + \frac{1}{\text{Re}}[\omega'' - k^2\omega] \qquad (10\text{-}18c)$$

$$\Phi' + ik\omega = 0 \qquad (10\text{-}18d)$$

The prime denotes differentiation with respect to x. The boundary conditions from Eq. (10-16a) are

$$\Phi = \psi = \omega = 0 \qquad \text{at } x = 0, 1 \tag{10-19}$$

The system is simplified by using Eq. (10-18d) to eliminate ω from Eq. (10-18c), and then differentiating Eq. (10-18c) once and combining with Eq. (10-18a) to eliminate η. The resulting system can be written as

$$\Phi^{IV} - \Phi'' - k^2(\Phi'' - k^2\Phi) = \lambda \, \text{Re} \left[\Phi'' - k^2\Phi + \frac{2k^2}{\lambda} \frac{\delta}{R}(1 - Mx)\psi \right] \tag{10-20a}$$

$$\psi'' - k^2\psi = \lambda \, \text{Re} \left[\psi - \frac{M}{\lambda} \Phi \right] \tag{10-20b}$$

$$\Phi = \Phi' = \psi = 0 \qquad \text{at } z = 0, 1 \tag{10-21}$$

A single sixth-order equation in Φ can be derived, but it is not helpful to do so, since approximation methods which require matching six boundary conditions would be prohibitive.

10.3 Inviscid Fluid

It is useful to examine first the case of the inviscid fluid, $\text{Re} \to \infty$. In that case the terms multiplying Re in Eqs. (10-20) must go to zero, and we obtain equations

$$\Phi'' - k^2\Phi + \frac{2k^2}{\lambda} \frac{\delta}{R}(1 - Mx)\psi = 0 \tag{10-22a}$$

$$\psi - \frac{M}{\lambda} \Phi = 0 \tag{10-22b}$$

This limit is singular, in that the order of the Φ equation falls from fourth to second and the ψ equation from second to zero. Compatibility therefore requires that the boundary condition on Φ' be dropped so that the four remaining conditions can be satisfied. Combining Eqs. (10-22) leads to a single second-order equation

$$\Phi'' + \left[-k^2 + \frac{2k^2}{\lambda^2} \frac{M\delta}{R}(1 - Mx) \right]\Phi = 0 \tag{10-23}$$

$$\Phi(0) = \Phi(1) = 0 \tag{10-24}$$

The δ/R term is the Coriolis force.

Equation (10-23) is of the form considered in Appendix D. By slightly modifying the proof in Sec. D.3 it can be shown that λ^2 is real. From inequality (D.7) it follows that a nontrivial solution exists only if

$$-k^2 + \max \frac{2k^2 M\delta}{\lambda^2 R} (1 - Mx) \geq \pi^2 \qquad (10\text{-}25)$$

Two cases need to be considered. When M is negative, inequality (10-25) can be satisfied only for $\lambda^2 < 0$. On the other hand, when M is positive, then for each $k^2 > 0$ there is some $\lambda^2 > 0$ such that the inequality is satisfied. $\lambda^2 < 0$ means that λ is imaginary and the solutions in Eqs. (10-17) do not grow without bound. Thus, $M < 0$ is a stable configuration. (Note that the disturbance cannot actually die out, because the only mechanism for damping is viscosity, but we consider an inviscid fluid to be stable when a disturbance cannot grow.) $\lambda^2 > 0$ means that there is a root $\lambda > 0$, corresponding to a growing exponential in Eq. (10-17). For an inviscid fluid, therefore, the necessary and sufficient condition for stability is $M < 0$, or

$$\Omega_2 > \Omega_1 \qquad (10\text{-}26)$$

Equation (10-26), which states that a stable configuration requires that the outer cylinder rotate faster than the inner cylinder and in the same direction, is know as the *Rayleigh criterion*. Rayleigh deduced this result from physical considerations based on the conservation of angular momentum. We would expect the damping effect of viscosity to lead to a relaxation of the condition for stability, for we know, for example, that a stable flow exists at least at low rotational speeds for $\Omega_2 = 0$, $\Omega_1 \neq 0$.

10.4 Viscous Liquid

For a viscous liquid with Re finite it is convenient to define the new variable

$$\phi = M \text{ Re } \Phi \qquad (10\text{-}27)$$

and the dimensionless *Taylor number*,

$$T = \frac{2M\delta}{R} \text{Re}^2 = \frac{2\Omega_1[\Omega_1 - \Omega_2]\delta^3 R\rho^2}{\mu^2} \qquad (10\text{-}28)$$

Equations (10-20) and (10-21) are then written as

$$\phi^{IV} - 2k^2\phi'' + k^4\phi - \lambda \text{ Re } [\phi'' - k^2\phi] = k^2 T[1 - Mx]\psi \qquad (10\text{-}29a)$$

$$\psi'' - k^2\psi - \lambda \operatorname{Re} \psi = -\phi \qquad (10\text{-}29\text{b})$$

$$\phi = \phi' = \psi = 0 \qquad \text{at } x = 0, 1 \quad (10\text{-}30)$$

Note that the term involving the Taylor number arises from the Coriolis term.

The mathematical problem that we face here is not very different from that for the catalyst particle in Chapter 9. There is a physical difference, however, which slightly influences the manner of solution. For the catalyst the parameters are all physicochemical quantities which cannot be easily manipulated in the course of a single experiment. Thus, the steady-state solution is either stable or unstable, but we cannot perform an experiment in which we bring about a transition from a stable to an unstable state. On the other hand, we can do exactly that for the Couette flow by slowly increasing the rotational speed. Hence, while for the catalyst particle we wish to determine for a particular steady state whether the real part of the eigenvalue is positive or negative, in the Couette flow we are more interested in determining the rotational speed at which the real part of λ changes from negative to positive. That is, we are looking for the point of *marginal stability* where the real part of λ passes through zero. (Compare Sec. 4.5.)

For the special case $M = 0$ it can be shown using a proof like that in Sec. D.3 that λ is real, and λ was found to be real for an unstable configuration ($\lambda^2 > 0$) for the inviscid fluid. In general, however, a direct proof that λ is real cannot be carried out. Nevertheless, in seeking a solution we shall *assume* that λ is real. It was noted in Sec. 4.5 that this assumption of a real eigenvalue is often known as the *principle of exchange of stabilities*. This assumption is not necessary, but it greatly simplifies the calculations. We shall return to the point subsequently. If λ is real and we are looking for the point of marginal stability, then we are interested in solutions of Eq. (10-29) when $\lambda = 0$,

$$\phi^{\mathrm{IV}} - 2k^2\phi'' + k^4\phi = k^2 T[1 - Mx]\psi \qquad (10\text{-}31\text{a})$$

$$\psi'' - k^2\psi = -\phi \qquad (10\text{-}31\text{b})$$

The Taylor number is now the only physical parameter which appears, and, as emphasized on several occasions, the right-hand side of Eq. (10-31a) arises from the Coriolis force.

Solution of Eqs. (10-30) and (10-31) is most easily carried out using the partial integration approach to Galerkin's method outlined in Sec. 8.5. We write

$$\psi(x) = \sum_{n=1}^{N} C_n \sin n\pi x \qquad (10\text{-}32)$$

This choice of approximating function satisfies both boundary conditions on ψ, and, since the sine functions form a complete set, we obtain an exact solution for $N \to \infty$. (The solution method is, in fact, an application of the finite Fourier sine transform when $N \to \infty$.) Substituting Eq. (10-32) into the right-hand side of Eq. (10-31a) we can solve for $\phi(x)$ (using, for example, the method of undetermined coefficients),

$$\phi(x) = Tk^2 \sum_{n=1}^{N} \frac{C_n}{S_n^2} \left\{ [1 - Mx] \sin n\pi x - \frac{4n\pi M}{S_n} \cos n\pi x \right.$$

$$\left. + [P_n + Q_n x] \sinh kx + \left[\frac{4n\pi M}{S_n} - n\pi x - kP_n x \right] \cosh kx \right\} \qquad (10\text{-}33)$$

$$S_n = n^2\pi^2 + k^2 \qquad (10\text{-}34a)$$

$$P_n = \frac{1}{\sinh^2 k - k^2} \left\{ n\pi k - \frac{4n\pi M}{S_n} [k + \cosh k \sinh k - \{-1\}^n \{k \cosh k \right.$$

$$\left. + \sinh k\}] + [-1]^n [1 - M] n\pi \sinh k \right\} \qquad (10\text{-}34b)$$

$$Q_n = \frac{1}{\sinh^2 k - k^2} \left\{ n\pi [\sinh k \cosh k - k] - \frac{4n\pi M}{S_n} k \sinh k [\sinh k \right.$$

$$\left. - \{-1\}^n k] - [-1]^n [1 - M] n\pi [\sinh k - k \cosh k] \right\} \qquad (10\text{-}34c)$$

Equations (10-32) and (10-33) for ψ and ϕ are now substituted into Eq. (10-31b), and the residual is multiplied by $\sin mkx$, integrated from zero to unity, and set equal to zero. This leads to the homogeneous equations

$$\sum_{n=1}^{N} S_n [Tk^2 W_{mn} - \delta_{mn}] C_n = 0, \qquad m = 1, 2, \ldots, N \qquad (10\text{-}35)$$

$$\delta_{mn} = \begin{cases} +1, & m = n \\ 0, & m \neq n \end{cases}$$

$$W_{mn} = \frac{2 - M}{2S_n^3} - \frac{8mn\pi^2 M}{S_m S_n^4} \{1 - [-1]^m \cosh k\} + \frac{2m\pi[-1]^m}{S_m S_n^3} [n\pi + kP_n] \cosh k$$

$$- \frac{2m\pi}{S_m S_n^3} [-1]^m [P_n + 2Q_n] \sinh k$$

$$- \frac{4km\pi}{S_m^2 S_n^3} \{Q_n[1 - \{-1\}^m \cosh k] + [-1]^m k[n + kP_n] \sinh k\} + U_{mn}$$

$$U_{mn} = \begin{cases} 0, & m + n \text{ even} \\ \dfrac{16mn\pi M}{S_n^3 \pi [m^2 - n^2]} \left\{ \dfrac{2}{\pi^2 [m^2 - n^2]^2} - \dfrac{1}{S_n} \right\}, & m + n \text{ odd} \end{cases}$$

The Taylor number is then obtained as an eigenvalue by setting the determinant of the coefficient matrix to zero.

The structure is essentially revealed by taking $N = 1$. Setting the coefficient of C_1 to zero gives

$$T = \frac{1}{k^2 W_{11}} = \frac{2}{2 - M} \frac{[\pi^2 + k^2]^5 [k + \sinh k]}{k^2 \{[\pi^2 + k^2]^2 [k + \sinh k] - 16k\pi^2 \cosh^2 \frac{1}{2}k\}} \qquad (10\text{-}36)$$

Figure 10.2 shows a plot of $[2 - M]T$ versus k. For each k there is some value of T which represents a point of marginal stability, beyond which λ will have a positive real part and a disturbance will grow. We shall be summing separation-of-variables solutions (10-17) over all k to obtain the complete solution

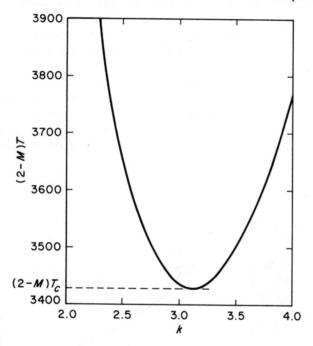

Figure 10.2 Curve of marginal stability for flow between rotating cylinders using one term in Galerkin's method.

to the partial differential equations (10-14), so the solution will contain a growing mode as soon as T reaches the minimum point on the T-k curve. This critical value, at which flow transition will occur, is at $k_c \simeq 3.12$, giving

$$T_c = \frac{3430}{2 - M} \qquad (10\text{-}37)$$

The flow is computed to be unstable for $T > T_c$. Better accuracy is obtained, of course, by using more terms in the expansion, but the result for $N = 1$ is surprisingly accurate for the first eigenvalue in the range $1 \geq \Omega_2/\Omega_1 \geq -0.5$. When $M = 1$ (inner cylinder rotating, outer stationary), for example, the critical Taylor number using three terms changes only from 3430 to 3390, and the critical wave number, k_c, is unchanged. Other eigenvalues are at larger values of T and are not of physical interest. The coefficients in the expansion (10-32) can be evaluated once the eigenvalues have been determined. For later reference we note the values here for $\Omega_2 = 0$ $(M = 1)$ at the critical point:

$$C_2 = 0.0039C_1, \qquad C_3 = -0.0011C_1$$

Clearly, the higher harmonics contribute negligibly to the solution for $M = 1$.

Figures 10.3 and 10.4 are plots of critical Taylor and wave number versus Ω_2/Ω_1 for $\Omega_2/\Omega_1 > -0.73$ with $N = 1$ and $N = 3$. These results have been verified experimentally at various Ω_2/Ω_1 in small gap instruments $(R/\delta \sim 20\text{–}40)$, and *any deviation between predicted $(N = 3)$ and observed transition in the most precisely carried out experiments is too small to be shown on the scale in Fig. 10.3.* When $\Omega_2/\Omega_1 \leq -0.73$ the assumption of axisymmetry is no longer valid, and the analysis given here does not predict the transition properly. As sketched in Fig. 1.6, the flow which follows transition consists of regularly spaced two-dimensional cells superimposed on the basic motion.

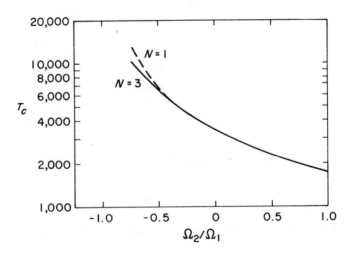

Figure 10.3 Critical Taylor number as a function of relative speeds of the cylinders, one and three terms in Galerkin's method.

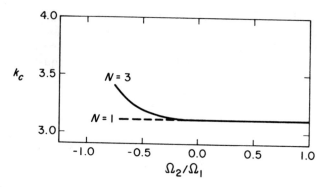

Figure 10.4 Critical wave number as a function of relative speeds of the cylinders, one and three terms in Galerkin's method.

The term exp (ikz) in Eqs. (10-17) corresponds to a periodic motion with dimensionless wavelength

$$\text{wavelength} = \frac{2\pi}{k}$$

When $k = k_c = 3.12$ for $M = 1$ the wavelength is almost exactly two dimensionless units, or twice the gap, δ. Agreement between the critical wavelength and experimental cell size is also excellent.

A word of caution is in order here. Agreement between the experimental transition and the prediction is not surprising. An earlier transition could possibly occur, since we have not checked for solutions which are non-axisymmetric or which have complex eigenvalues. Neither situation is ever seen experimentally, however; and the solution with complex eigenvalues has been carried out and verifies the validity of the assumption of exchange of stabilities. It is also possible that an earlier transition could occur because of finite disturbances not accounted for in a linear analysis. That does not seem to be the case, and linear theory accurately predicts the transition. The cellular flow field following transition, however, is a finite-amplitude flow which is beyond the range of the linear analysis. Thus, there is no reason to expect the linear theory ever to predict the cell spacing accurately, though in fact it appears to do so in this problem. We shall return to this point briefly in Chapter 16.

10.5 Non-Newtonian Liquid

In a non-Newtonian liquid, such as a polymer solution, the stress is not related to deformation rate through Eq. (10-3). The Taylor stability problem can be solved for a very general fluid in which the stress is taken as depending on the entire past history of the deformation gradient. This formulation is too general for comparison with experiment, however. If we use the simplest stress-deformation relation which completely describes the stress distribution in the Couette flow prior to instability, the linearized equations equivalent to Eqs. (10-31) become

$$\phi^{IV} - 2k^2\phi'' + k^4\phi = Tk^2\Bigg\{[1 - Mx]\psi -$$

$$-\frac{R}{\delta}\frac{\tilde{\tau}_{rr} - \tilde{\tau}_{\theta\theta}}{2\rho\,\delta^2\Gamma^2}\Bigg\{\bigg[1 - \frac{d\ln\tilde{\tau}_{rr} - \tilde{\tau}_{\theta\theta}}{d\ln\Gamma}\bigg]\psi'' - k^2\psi\bigg\}\Bigg\} \quad (10\text{-}38a)$$

$$\bigg[1 + \frac{d\ln\mu}{d\ln\Gamma}\bigg]\psi'' - k^2\psi = -\phi + \Bigg\{\frac{\tilde{\tau}_{\theta\theta} - \tilde{\tau}_{zz} + 2[\tilde{\tau}_{rr} - \tilde{\tau}_{\theta\theta}]}{2\rho\,\delta^2\Gamma^2}\Bigg\}\{\psi'' - k^2\psi\}$$

$$(10\text{-}38b)$$

The viscosity, μ, is in general a function of shear rate, Γ. T is calculated using the viscosity corresponding to the critical conditions. The stress differences $\tilde{\tau}_{rr} - \tilde{\tau}_{\theta\theta}$ and $\tilde{\tau}_{\theta\theta} - \tilde{\tau}_{zz}$ refer to conditions in the undisturbed flow. Their dependence on shear rate can be determined independently on an instrument known as a *rheogoniometer*, though $\tilde{\tau}_{rr} - \tilde{\tau}_{zz}$ is small and very difficult to measure accurately in most polymer solutions. For a Newtonian fluid, $\mu = $ constant, $\tilde{\tau}_{rr} = \tilde{\tau}_{\theta\theta} = \tilde{\tau}_{zz}$, and we recover Eqs. (10-31).

The structure of Eqs. (10-38) is identical to that of Eqs. (10-31), and the same solution method can be applied without any additional complexity. In general it is found that the critical Taylor number is predicted to be increased relative to a Newtonian liquid, while the critical wave number is decreased. Very few experiments have been carried out. For a solution of 0.14% polyacrylamide, 6.29% water, and 93.57% glycerine the functions in Eqs. (10-38) are given by

$$\mu = 4.35\Gamma^{-0.15} \qquad \tilde{\tau}_{\theta\theta} - \tilde{\tau}_{rr} = 7.35\Gamma^{1.05} \qquad \tilde{\tau}_{rr} - \tilde{\tau}_{zz} = -22.20\Gamma^{0.47}$$

Stresses are in dynes per square centimeter, and Γ is in reciprocal seconds. There is considerable uncertainty in $\tilde{\tau}_{rr} - \tilde{\tau}_{zz}$. In a system with $\delta/R = 0.031$ the critical Taylor number is predicted to be $T_c = 4330$, with confidence limits $3870 \le T_c \le 4660$ because of the uncertainity in $\tilde{\tau}_{rr} - \tilde{\tau}_{zz}$. The experimental transition occurred at $\Gamma = 3800$ sec^{-1}, $T_c = 4060$, which is good agreement. The predicted wave number is $k_c = 2.57$, or, with confidence limits, $2.78 \ge k_c \ge 2.45$. The experimental wave number was 3.22, which is considerably different. We shall return to this discrepancy between experimental and linear theoretical wave number in Chapter 16, where we shall develop a nonlinear analysis similar to that in Sec. 6.2. Based on that analysis it can be established that the minimum energy configuration for the finite-amplitude flow of a non-Newtonian liquid requires a higher wave number than the linear value, in accordance with experiment.

A viscoelastic non-Newtonian liquid has energy storage mechanisms which do not exist for a Newtonian liquid, and one might suspect the possibility of an elasticity-induced oscillatory, or *overstable*, response, corresponding to a complex eigenvalue, λ. Under certain conditions such behavior is predicted, and an oscillatory instability has been observed in moderately concentrated polymer solutions.

BIBLIOGRAPHICAL NOTES

The material in this chapter, with extensions and more general flows, including hydromagnetic effects, is covered elegantly and completely in

> Chandrasekhar, S., *Hydrodynamic and Hydromagnetic Stability*, Oxford University Press, Inc., New York, 1961.

The basic study is by Taylor,

> Taylor, G. I., *Phil. Trans. Royal Soc.*, *A223*, 289 (1923).

See also Stuart's review paper,

> Stuart, J. T., *Appl. Mech. Rev.*, *18*, 523 (1965).

The principle of exchange of stabilities is discussed and proved for $\Omega_2/\Omega_1 > 0$ in

> Yih, C. S., *Arch. Rat. Mech. Anal.*, *46*, 218 (1972); *47*, 288 (1972).

The demonstration that the assumption of axisymmetry breaks down for $\Omega_2/\Omega_1 < -0.73$ and the solution for nonaxisymmetric modes is in

> Krueger, E. R., A. Gross, and R. C. DiPrima, *J. Fluid Mech.*, *24*, 521 (1966).

For a detailed experimental study of the formation of Taylor vortices and their behavior at increasing rotational speeds, see

> Burkhalter, J. E., and E. L. Koschmieder, *J. Fluid Mech.*, *58*, 547 (1973).
> Coles, D., *J. Fluid Mech.*, *21*, 385 (1965).
> Snyder, H., *Phys. Fluids*, *11*, 728 (1968).

An extension of the analysis to eccentric cylinders, of interest in applications of journal bearings, is in

> DiPrima, R. C., and J. T. Stuart, *J. Fluid Mech.*, *54*, 393 (1972).

The material on non-Newtonian fluids in Sec. 10-5 is based on

> Sun, Z.-S., and M. M. Denn, *AIChE J.* *18*, 1010 (1972).

where complete references to earlier work by H. Giesekus, K. Walters, and others may be found. Additional experiments are reported in

Jones, W. M., D. M. Davies, and M. C. Thomas, *J. Fluid Mech.*, *60*, 19 (1973).

Linear Stability: Plane Poiseuille Flow

11

11.1 Introduction

The transition between laminar and turbulent flow in a straight conduit is extremely important in engineering design. Laminar flow, with straight streamlines, requires less pumping energy than the chaotic turbulent motion, whereas turbulence is required for effective mixing in a reactor or heat exchanger. The transition is better understood in flow between parallel plates than in flow in a cylinder, and we shall consider only the planar geometry here.

The flow geometry is shown in Fig. 11.1. Flow is in the x direction because of a constant pressure gradient, $\partial p/\partial x$. The plates are separated by a

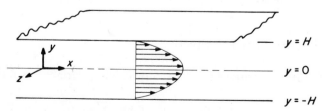

Figure 11.1 Schematic diagram for flow between flat plates.

distance $2H$ in the y direction. There are no body forces. In the rectangular Cartesian coordinates the dimensionless Navier-Stokes and continuity equations, (10-5) and (10-6), are

$$\frac{\partial v_x}{\partial t} + v_x \frac{\partial v_x}{\partial x} + v_y \frac{\partial v_x}{\partial y} + v_z \frac{\partial v_x}{\partial z} = -\frac{\partial p}{\partial x} + \frac{1}{Re}\left[\frac{\partial^2 v_x}{\partial x^2} + \frac{\partial^2 v_x}{\partial y^2} + \frac{\partial^2 v_x}{\partial z^2}\right] \quad (11\text{-}1a)$$

$$\frac{\partial v_y}{\partial t} + v_x \frac{\partial v_y}{\partial x} + v_y \frac{\partial v_y}{\partial y} + v_z \frac{\partial v_y}{\partial z} = -\frac{\partial p}{\partial y} + \frac{1}{Re}\left[\frac{\partial^2 v_y}{\partial x^2} + \frac{\partial^2 v_y}{\partial y^2} + \frac{\partial^2 v_y}{\partial z^2}\right] \quad (11\text{-}1b)$$

$$\frac{\partial v_z}{\partial t} + v_x \frac{\partial v_z}{\partial x} + v_y \frac{\partial v_z}{\partial y} + v_z \frac{\partial v_z}{\partial z} = -\frac{\partial p}{\partial z} + \frac{1}{Re}\left[\frac{\partial^2 v_z}{\partial x^2} + \frac{\partial^2 v_z}{\partial y^2} + \frac{\partial^2 v_z}{\partial z^2}\right] \quad (11\text{-}1c)$$

$$\frac{\partial v_x}{\partial x} + \frac{\partial v_y}{\partial y} + \frac{\partial v_z}{\partial z} = 0 \quad (11\text{-}2)$$

Boundary conditions are no-slip at the solid surfaces,

$$v_x = v_y = v_z = 0 \qquad \text{at } y = \pm 1 \quad (11\text{-}3a)$$

$$v_x, v_y, v_z \text{ bounded as } x, z \rightarrow \pm \infty \quad (11\text{-}3b)$$

Here, the distance coordinate has been made dimensionless with respect to the half-width, H, and the reference velocity, U, is taken as the *maximum* velocity. (The use of the maximum velocity is unusual, but it simplifies some of the numerical factors which are carried in the stability analysis.) The steady-state solution is the parabolic velocity profile of laminar, or plane Poiseuille flow,

$$v_{xs} = 1 - y^2, \qquad v_{ys} = v_{zs} = 0 \quad (11\text{-}4)$$

$$p_s = p_0 + \frac{\partial p}{\partial x} x \quad (11\text{-}5)$$

The solution is independent of the Reynolds number and is observed experimentally up to $Re \simeq 1000$. The Reynolds number defined on the basis of half-width and maximum velocity is $\frac{3}{4}$ the Reynolds number defined with the mean velocity and full channel width.

11.2 Perturbation Equations

The perturbation equations are derived by defining variables as follows:

$$u = v_x - v_{xs} \qquad v = v_y - v_{ys}$$

$$w = v_z - v_{zs} \qquad q = p - p_s$$

Upon substitution into the Navier-Stokes and continuity equations, (11-1) and (11-2), and neglecting second-order terms, we obtain the following set of linear partial differential equations:

$$\frac{\partial u}{\partial t} + [1 - y^2]\frac{\partial u}{\partial x} - 2yv = -\frac{\partial q}{\partial x} + \frac{1}{Re}\left[\frac{\partial^2 u}{\partial x^2} + \frac{\partial^2 u}{\partial y^2} + \frac{\partial^2 u}{\partial z^2}\right] \qquad (11\text{-}6a)$$

$$\frac{\partial v}{\partial t} + [1 - y^2]\frac{\partial v}{\partial x} = -\frac{\partial q}{\partial y} + \frac{1}{Re}\left[\frac{\partial^2 v}{\partial x^2} + \frac{\partial^2 v}{\partial y^2} + \frac{\partial^2 v}{\partial z^2}\right] \qquad (11\text{-}6b)$$

$$\frac{\partial w}{\partial t} + [1 - y^2]\frac{\partial w}{\partial x} = -\frac{\partial q}{\partial z} + \frac{1}{Re}\left[\frac{\partial^2 w}{\partial x^2} + \frac{\partial^2 w}{\partial y^2} + \frac{\partial^2 w}{\partial z^2}\right] \qquad (11\text{-}6c)$$

$$\frac{\partial u}{\partial x} + \frac{\partial v}{\partial y} + \frac{\partial w}{\partial z} = 0 \qquad (11\text{-}7)$$

The boundary conditions follow from Eqs. (11-3) as

$$u = v = w = 0 \qquad \text{at } y = \pm 1 \qquad (11\text{-}8a)$$

$$u, v, w = \text{bounded as } x, z \to \pm \infty \qquad (11\text{-}8b)$$

Equations (11-6) and (11-7) contain *four* independent variables, x, y, z, and t. If we seek a solution by separation of variables, we find that the velocity must be a sum or integral of terms of the form

$$u(x, y, z, t) = e^{\lambda t}e^{\beta x}e^{\gamma z}\phi(y)$$

The condition of boundedness at $x, z = \pm \infty$ requires that β and γ be imaginary; we write $\beta = ik_x$, $\gamma = ik_z$, where k_x and k_z are positive real numbers and denote Fourier wave numbers in the x and z directions, respectively. It is customary to write

$$\lambda = -ik_x c$$

c is a complex dimensionless velocity,

$$c = c_R + ic_I$$

A disturbance which grows corresponds to $c_I > 0$, one which decays to $c_I < 0$, and marginal stability to $c_I = 0$. With this notation it is readily established that the solution by separation of variables for all terms is of the form

$$u = e^{-ik_x ct}e^{ik_x x}e^{ik_z z}\phi(y) \qquad (11\text{-}9a)$$

$$v = e^{-ik_x ct}e^{ik_x x}e^{ik_z z}\psi(y) \qquad (11\text{-}9b)$$

$$w = e^{-ik_x ct} e^{ik_x x} e^{ik_z z} \omega(y) \tag{11-9c}$$

$$q = e^{-ik_x ct} e^{ik_x x} e^{ik_z z} \eta(y) \tag{11-9d}$$

When these relations are substituted into Eqs. (11-6) and (11-7) we obtain

$$\phi'' - [k_x^2 + k_z^2 + ik_x \, \text{Re} \, (1 - y^2 - c)]\phi = -2y \, \text{Re} \, \psi + ik_x \, \text{Re} \, \eta \tag{11-10a}$$

$$\psi'' - [k_x^2 + k_z^2 + ik_x \, \text{Re} \, (1 - y^2 - c)]\psi = \text{Re} \, \eta' \tag{11-10b}$$

$$\omega'' - [k_x^2 + k_z^2 + ik_x \, \text{Re} \, (1 - y^2 - c)]\omega = ik_z \, \text{Re} \, \eta \tag{11-10c}$$

$$ik_x \phi + ik_z \omega + \psi' = 0 \tag{11-11}$$

The prime denotes differentiation with respect to y. Boundary conditions from Eq. (11-8a) are

$$\phi(-1) = \phi(1) = \psi(-1) = \psi(1) = \omega(-1) = \omega(1) = 0 \tag{11-12}$$

We shall show in the next section that the most unstable disturbance is a two-dimensional disturbance in the x-y plane with $\omega = 0$, $k_z = 0$. In that case, setting $k_x = k$, Eqs. (11-10) and (11-11) simplify to

$$\phi'' - [k^2 + ik \, \text{Re} \, (1 - y^2 - c)]\phi = -2y \, \text{Re} \, \psi + ik \, \text{Re} \, \eta \tag{11-13a}$$

$$\psi'' - [k^2 + ik \, \text{Re} \, (1 - y^2 - c)]\psi = \text{Re} \, \eta' \tag{11-13b}$$

$$ik\phi + \psi' = 0 \tag{11-14}$$

These can be combined to a single fourth-order equation for ψ,

$$\psi^{\text{IV}} - 2k^2 \psi'' + k^4 \psi - ik \, \text{Re} \, [(1 - y^2 - c)(\psi'' - k^2 \psi) + 2\psi] = 0 \tag{11-15}$$

Equation (11-15) is known as the *Orr-Sommerfeld equation*. The two additional boundary conditions for ψ are obtained by substituting Eq. (11-12) into (11-14) at $y = \pm 1$, giving

$$\psi(1) = \psi(-1) = \psi'(1) = \psi'(-1) = 0 \tag{11-16}$$

11.3 Squire's Theorem

We shall prove that the two-dimensional disturbance is the most unstable. This is done by showing that for every three-dimensional disturbance which grows in time there is a corresponding two-dimensional disturbance which also grows, and the two-dimensional disturbance becomes unstable at a lower Reynolds number than the three-dimensional disturbance. The result is known as *Squire's theorem*.

The proof is carried out by introducing new variables into Eqs. (11-10) and (11-11) as follows:

$$\phi_* = \frac{k_x \phi + k_z \omega}{k} \tag{11-17a}$$

$$\eta_* = \frac{\eta k_x}{k} \tag{11-17b}$$

$$\text{Re}_* = \frac{\text{Re}\, k_x}{k} \tag{11-17c}$$

where

$$k^2 = k_x^2 + k_z^2 \tag{11-17d}$$

Note that

$$\text{Re}_* \leq \text{Re}$$

Substituting these new variables into Eq. (11-10a) gives

$$k\{\phi_*'' - [k^2 + ik\,\text{Re}_* (1 - y^2 - c)]\phi_*\} = -2yk\,\text{Re}_*\,\psi + ik_x^2\,\text{Re}_*\,\eta_*$$
$$+ k_z\{\omega'' - [k^2 + ik\,\text{Re}_* (1 - y^2 - c)]\omega\} \tag{11-18}$$

The bracketed term involving ω is replaced using Eq. (11-10c), leading finally to

$$\phi_*'' - [k^2 + ik\,\text{Re}_* (1 - y^2 - c)]\phi_* = -2y\,\text{Re}_*\,\psi + ik\,\text{Re}_*\,\eta_* \tag{11-19a}$$

In a similar way we obtain from Eqs. (11-10b) and (11-11) the equations

$$\psi'' - [k^2 + ik\,\text{Re}_* (1 - y^2 - c)]\psi = \text{Re}_*\,\eta_*' \tag{11-19b}$$

$$ik\phi_* + \psi' = 0 \tag{11-20}$$

These combine to the single fourth-order equation for ψ,

$$\psi^{\text{IV}} - 2k^2\psi'' + k^4\psi - ik\,\text{Re}_* [(1 - y^2 - c)(\psi'' - k^2\psi) + 2\psi] = 0 \tag{11-21}$$

This is identical to the Orr-Sommerfeld equation (11-15), and the boundary conditions are readily shown to be those in Eq. (11-16). Thus, we can conclude that whenever there is a three-dimensional disturbance with growth rate $k_x c_I$ at Reynolds number Re, there is a two-dimensional disturbance with growth rate kc_I at a *lower* Reynolds number, Re_*. Any time c_I is positive, corresponding to a growing disturbance, the lowest Reynolds number for which that positive value of c_I will occur is at $k_z = 0$, $\omega = 0$.

11.4 Inviscid Fluid

Some useful insight into the nature of the solution can be obtained by considering the limit of an inviscid fluid. If $Re \to \infty$ in Eq. (11-15), the quantity in the brackets multiplying Re must go to zero. Thus, we obtain the *inviscid Orr-Sommerfeld equation,*

$$(1 - y^2 - c)(\psi'' - k^2\psi) + 2\psi = 0 \qquad (11\text{-}22)$$

The boundary conditions retained in this limit are

$$\psi(1) = \psi(-1) = 0 \qquad (11\text{-}23)$$

These correspond to a vanishing velocity normal to the solid surfaces. The conditions $\psi'(1) = \psi'(-1) = 0$ correspond to no-slip tangent to the solid surfaces and cannot be satisfied by an inviscid fluid.

We shall focus attention on disturbances which are either growing or decaying in time. In that case $c_I \neq 0$ and $(1 - y^2 - c)$ does not vanish in the interval $-1 \leq y \leq 1$. It is convenient to change to the new variable

$$\xi(y) = \frac{\psi(y)}{1 - y^2 - c} \qquad (11\text{-}24)$$

Note that $\xi(y)$ is a complex function,

$$\xi(y) = \xi_R(y) + i\xi_I(y)$$

Equations (11-22) and (11-23) then become

$$\frac{d}{dy}[(1 - y^2 - c)\xi'] - k^2(1 - y^2 - c)\xi = 0 \qquad (11\text{-}25)$$

$$\xi(-1) = \xi(1) = 0 \qquad (11\text{-}26)$$

The following steps, similar to those in Appendix D.2, are then carried out:

1. Equation (11-25) is separated into real and imaginary parts.

2. The equation for ξ_R is multiplied by ξ_I, and the equation for ξ_I is multiplied by ξ_R.
3. Both equations are integrated from -1 to $+1$ and added.

The result is

$$c_I \int_{-1}^{1} (1 - y^2 - c_R)[\xi_R'^2 + \xi_I'^2 + k^2(\xi_R^2 + \xi_I^2)] \, dy = 0 \qquad (11\text{-}27)$$

This equation can be satisfied only if $1 - y^2 - c_R$ changes sign in the interval $-1 \le y \le 1$, which in turn implies that

$$0 < c_R < 1 \qquad (11\text{-}28)$$

Thus, at least in the inviscid limit, all disturbances will contain a non-vanishing term $\exp(-ikc_R t)$ and will grow or decay in an oscillatory manner. The principle of exchange of stabilities is not satisfied, in contrast to the behavior of disturbances in the Taylor problem studied in the preceding chapter.

11.5 Solution of the Orr-Sommerfeld Equation

The properties of the Orr-Sommerfeld equation have been studied extensively. Analytical solution is difficult because of some theoretical problems associated with the choice of branches in the complex plane and with behavior in the *critical layer*. To explain the latter we need to note that the minimum of the curve of neutral stability will occur at $\mathrm{Re} > 5000$. In general, then, the inviscid limit will be approximately applicable. Near the neutral curve $c_I \simeq 0$, and we have shown that in the inviscid limit $0 < c_R < 1$. Thus, there is a small region, where $1 - y^2 \simeq c_R$, in which the large Reynolds number multiplies a small quantity, the higher derivatives become important, and the qualitative character of the equation changes significantly. This region, called the critical layer, manifests itself by rapid changes in the values of the eigenfunctions over small distances. For the same reason, conventional single-precision numerical integration methods cannot be used to solve the equation.

The eigenvalues of the Orr-Sommerfeld equation can be found using Galerkin's method. We assume that the solution is even about $y = 0$, in which case the expansion in an appropriate set of approximating functions which satisfies the four boundary conditions is

$$\psi(y) = \sum_{n=1}^{N} C_n(1 - y^2)^2 y^{2n-2} \qquad (11\text{-}29)$$

A one-term expansion reveals some structure. The algebraic eigenvalue equation is

$$12.80 + 2.44k^2 - 0.41ik \text{ Re} + 0.41k^4 + 0.37ik^3 \text{ Re}$$
$$- (c_R + ic_I)(0.41ik^3 \text{ Re} + 1.22ik \text{ Re}) = 0$$

This can be separated into real and imaginary parts to give the eigenvalue, $c = c_R + ic_I$,

$$c_R = \frac{1 + 0.9k^2}{3 + k^2} \tag{11-30a}$$

$$c_I = -\frac{12.80 + 2.44k^2 + 0.41k^4}{0.41k^3 \text{ Re} + 1.22k \text{ Re}} \tag{11-30b}$$

c_R is relatively insensitive and satisfies Eq. (11-28), $0 < c_R < 1$. c_I decreases in magnitude with increasing Re, though the one-term expansion is not adequate to locate a transition to positive c_I and instability.

As many as 13 terms are required in the expansion to obtain adequate convergence. Figure 11.2 shows the convergence of c_I for $k = 1$, Re $= 6400$,

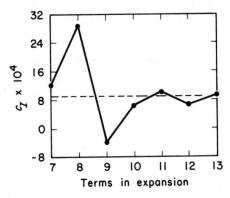

Figure 11.2 Convergence of the real part of the first eigenvalue of the Orr-Sommerfeld equation with terms in Galerkin's method, $k = 1$, Re $= 6400$.

with an increasing number of terms in the expansion. Four-decimal accuracy is required to ensure even the proper algebraic sign on c_I. Note that the convergence to the eigenvalue is oscillatory. This is to be expected in non-self-adjoint problems. Figure 11.3(a) and (b) shows c_R and c_I as functions of Re for $k = 1$. The point of marginal stability corresponds to points where c_I passes through zero. Thus, for $k = 1$, the flow is stable for Re < 5900 and Re $> 35,000$, and unstable in between.

The marginal or neutral stability curve, where $c_I = 0$, is plotted in Fig. 11.4. The minimum occurs at a Reynolds number of slightly less than 5800.

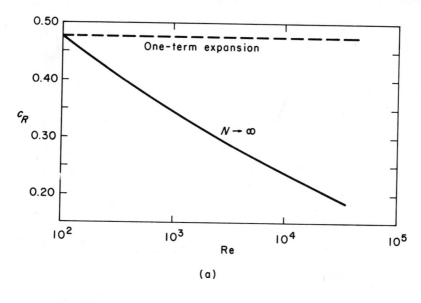

(a)

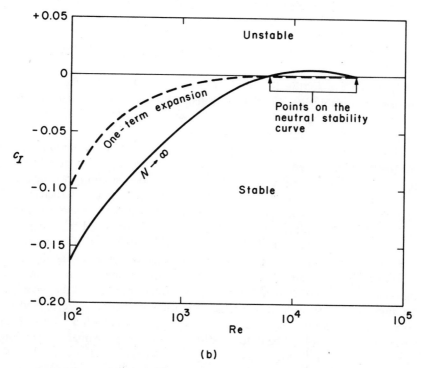

(b)

Figure 11.3 (a) Real and (b) imaginary parts of the first eigenvalue of the Orr-Sommerfeld equation for $k = 1$.

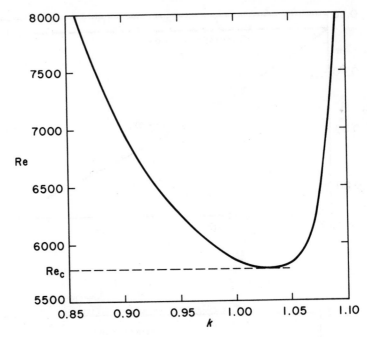

Figure 11.4 Curve of marginal stability for flow between flat plates.

Thus, according to the linear theory, there will be at least one term in the separation solution, Eq. (11-9), which will grow in time whenever Re > $Re_c \simeq 5800$, and beyond that point the parabolic laminar velocity distribution cannot be maintained.

Two key assumptions have been made in the solution. The first is that the disturbance is even about $y = 0$. This is not necessary but does simplify the computations, and it can be established that odd solutions to the eigenvalue equation are not destabilizing. The second is that we may focus on the first eigenvalue of the equation, which is the eigenvalue of smallest modulus ($[c_R^2 + c_I^2]^{1/2}$). The other eigenvalues can clearly be obtained from the same computer program used to generate the first, and they have been calculated at $k = 1$ and a number of values of Re. Good convergence of the first three eigenvalues is obtained. In all cases the eigenvalue of smallest modulus is the most unstable. The difference in modulus between successive eigenvalues can be quite small, however; so computer algorithms which iterate to find the eigenvalue of a matrix may reach the wrong eigenvalue with a poor starting value. Finally, it should be noted that the choice of approximating functions in Eq. (11-29) is not at all unique, or even obvious. L. H. Lee and W. C. Reynolds have studied the solution of the Orr-Sommerfeld equation using a variety of approximating functions. The final result is insensitive to the choice of functions, but the ones used here combined efficient convergence with extreme simplicity in carrying out the integrations required for Galerkin's method.

11.6 Dilute Polymer Solutions

The observation that trace amounts of certain water-soluble high-molecular-weight polymers can significantly reduce the frictional pressure drop in turbulent pipeline flow has prompted a number of investigations into the stability of laminar flow of these solutions. The viscoelastic properties can be characterized by a *relaxation time*, t_R, which is given by a molecular theory of polymer solutions as

$$t_R = \frac{6(\mu - \mu_s) M_w}{\pi^2 CRT} \qquad (11\text{-}31)$$

C is the polymer concentration, R the gas constant, M_w the molecular weight, T the absolute temperature, μ the solution viscosity, and μ_s the viscosity of the pure solvent. In dilute polymer solutions t_R is typically of order 10^{-4}–10^{-6} sec.

The stability analysis for such a liquid is carried out in a manner identical to the Newtonian fluid considered in this chapter, but now an extra parameter appears, the *elasticity number*,

$$E = \frac{t_R \mu}{\rho H^2} \qquad (11\text{-}32)$$

The generalization of the Orr-Sommerfeld equation is rather more complex in form and will not be repeated here. The eigenvalue, c, appears in the equation in a nonlinear fashion, so computation time is considerably increased. However, for k of order unity and Re E small compared to unity an approximation can be made which reduces the equation to one which differs only slightly from the Newtonian Orr-Sommerfeld equation,

$$[1 - ik \text{ Re } E(1 - y^2 - c)][\psi^{IV} - 2k^2\psi'' + k^4\psi]$$
$$= ik \text{ Re } [(1 - y^2 - c)(\psi'' - k^2\psi) + 2\psi] \quad (11\text{-}33)$$

Since Re will be of order 5000 in the region of the minimum of the neutral curve, E will have to be less than about 10^{-4} for this approximation to hold. $E = 10^{-4}$ is also a reasonable upper bound for flow of very dilute polymer solutions.

Figure 11.5 is a plot of critical Reynolds number versus elasticity number. The critical value computed from Eq. (11-33) agrees quite well with the result from the complete equation up to $E \simeq 5 \times 10^{-3}$. In the range $E \leq 5 \times 10^{-3}$ there is a slight decrease in Re_c with increasing E, but, as we shall note in the next section, the difference from Newtonian behavior is not significant. Beyond the dilute solution range, however, there is a dramatic

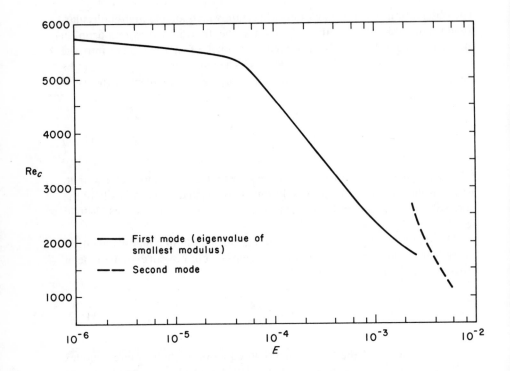

Figure 11.5 Critical Reynolds number as a function of Elasticity number for dilute polymer solutions. [After K. C. Porteous and M. M. Denn, *Trans. Soc. Rheol.*, *16*, 295 (1972), with permisssion.]

decrease which indicates a clear qualitative difference in behavior. Furthermore, for $E \simeq 10^{-3}$ the higher eigenvalues need to be considered, since they, too, have positive imaginary parts for Reynolds numbers which are comparable to the transition in the first mode. Beyond $E = 5 \times 10^{-3}$ the imaginary part of the eigenvalue of smallest modulus never becomes positive, and the transition is determined by an eigenvalue with a larger modulus (i.e., a higher mode). Thus, unlike the Newtonian liquid, all eigenvalues must be examined for the polymer solution, a process which is tedious and computationally expensive.

11.7 Comparison with Experiment

According to the computation in Fig. 11.4, the transition Reynolds number for a Newtonian fluid is about 5800. Experimentally, the laminar profile becomes turbulent at a Reynolds number around 1000. This transition appears to be governed by disturbances of finite amplitude. It is important to keep in mind that a linear stability analysis can only show when an infinitesimal disturbance will grow and thus provides an *upper bound* on the region of possible stability. A disturbance of finite amplitude might well grow at conditions under which an infinitesimal disturbance will decay. We shall show in Chapter 17 that such is indeed the case with plane Poiseuille flow.

These comments apply particularly to the viscoelastic liquid. The slight decrease in Re_c up to $E \simeq 10^{-4}$ is negligible compared to the fivefold effect of finite disturbances and cannot have physical significance. This observation is borne out experimentally, for differences in the turbulent transition between Newtonian liquids and dilute polymer solutions are not usually observed. At higher values of E, on the other hand, the linear transition is predicted to occur in a range where nonlinear effects dominate the Newtonian transition. At present there are no data available here.

Next, it should be noted that we have considered only flow between flat plates, while flow in a circular pipe is the problem of real pragmatic interest. All linear stability analyses carried out thus far for flow in a pipe show that the laminar flow is stable to two- and three-dimensional infinitesimal disturbances at all Reynolds numbers. Thus, the flat plate problem has generally been studied in the hope that it will provide some insight into laminar-turbulent transitions that can be applied more generally.

Finally, Eq. (11-15), derived in this chapter, is a special case of the slightly more general Orr-Sommerfeld equation which describes the stability of rectilinear and nearly rectilinear flows. Other configurations which have been treated include the laminar boundary layer and falling laminar film. Agreement with experiment is good for the boundary layer. For the falling film the effect of traces of surfactants on the surface tension is so strong that precise comparison with experiment is difficult. An analytical solution can be obtained for the falling film.

BIBLIOGRAPHICAL NOTES

The standard text on the stability of parallel flow is

Lin, C. C., *The Theory of Hydrodynamic Stability*, Cambridge University Press, New York, 1966.

Lin contains an extensive bibliography through 1955, the date of first printing. There is a detailed bibliography through 1966 in

Betchov, R., and W. O. Criminale, Jr., *Stability of Parallel Flows*, Academic Press, New York, 1967.

See also the survey papers

Reid, W. H., in M. Holt, ed., *Basic Developments in Fluid Mechanics*, Academic Press, New York, 1965.
Stuart, J. T., *Appl. Mech. Rev.*, *18*, 523 (1965).

There is a useful discussion of hydrodynamic stability in a chapter of

Monin, A. S., and A. M. Yaglom, *Statistical Fluid Mechanics*: *Mechanics of Turbulence*, vol.1, M.I.T. Press, Cambridge, Mass., 1971.

There is also a broad coverage in a chapter of

Yih, C.-S., *Fluid Mechanics*, McGraw-Hill, New York, 1969.

The properties of the inviscid Orr-Sommerfeld equation (Sec. 11.4) are described in detail in

Drazin, P. G., and L. N. Howard, *Adv. Appl. Mech.*, *9*, 1 (1966).

See also

Craik, A. D. D., *J. Fluid Mech.*, *53*, 657 (1972).

The numerical analysis of the Orr-Sommerfeld equation has received considerable attention. The most commonly used direct integration techniques are those of Kaplan and Nachtsheim:

Kaplan, R. E., "Solution of the Orr-Sommerfeld Equation for Laminar Boundary Layer Flow over Compliant Boundaries," *ASRL-TR-116-1*, Cambridge Aeroelastic and Structures Research Laboratory, M.I.T., Cambridge, Mass., 1964.
Nachtsheim, P. R., "An Initial Value Method for the Numerical Treatment of the Orr-Sommerfeld Equation for the Case of Plane Poiseuille Flow," *NASA Technical Note D-2414*, 1964.

Direct numerical procedures are discussed in some detail in Betchov and Criminale, cited above. A technique known as quasilinearization, which is the application of the Newton-Raphson method to functional equations, is exploited in

Radbill, J. R., and G. A. McCue, *Quasilinearization and Nonlinear Problems in Fluid and Orbital Mechanics*, American Elsevier, New York, 1970.

Lee and Reynolds have studied the effect of various approximating functions for Galerkin's method in a very detailed report,

Lee, L. H., and W. C. Reynolds, "A Variational Method for Investigating the Stability of Parallel Flows," *Technical Report FM-1*, Mechanical Engineering Department, Stanford University, Stanford, Calif., 1964.

The variational method of the title is identical to Galerkin's method for equations with the structure of the Orr-Sommerfeld equation, where the equation and its adjoint have the same boundary conditions. (See Appendix E.) Some results from the report, together with a discussion of Kaplan's method, have been published in

Lee, L. H., and W. C. Reynolds, *Quart. J. Mech. Appl. Math.*, 20, 1 (1967).

See also

Platten, J. K., *Intern. J. Eng. Sci.*, 9, 37 (1971).

Orszag, S. A., *J. Fluid Mech.*, 50, 689 (1971).

Gersting, J. M., Jr., and D. F. Jankowski, *Intern. J. Numerical Methods Eng.*, 4, 195 (1972).

The work of Lee and Reynolds and Orszag includes calculations of additional eigenvalues of the Orr-Sommerfeld equation. Experiments on transition in plane Poiseuille flow are reported in

Davies, S. J., and C. M. White, *Proc. Roy. Soc.*, 119, 92, (1928).

Kao, T. W., and C. Park, *J. Fluid Mech.*, 43, 145 (1970).

The effect of flexible walls on the stability of plane Poiseuille flow is treated in

Green, C. H., and C. H. Ellen, *J. Fluid Mech.*, 51, 403 (1972).

The discussion of flow stability of dilute polymer solutions in Sec. 11.6 is based on

Porteous, K. C., and M. M. Denn, *Trans. Soc. Rheol.*, 16, 295 (1972).

where references to earlier work may be found. Porteous's 1971 University of Deleware Ph.D. Thesis, cited in the paper, contains a detailed discussion of computation for both the Newtonian and slightly viscoelastic fluids. One important study not cited by Porteous and Denn is

Cousins, R. R., *Intern. J. Eng. Sci.*, 8, 595 (1970).

Cousins shows by calculation that Squire's theorem is not generally true for viscoelastic liquids. His calculations can be used to demonstrate, however, that the two-dimensional disturbance is the most unstable for model liquids whose physical properties correspond to those always observed experimentally.

Stability of laminar Newtonian flow in a round pipe has been studied by a number of workers and always found to be stable to infinitesimal disturbances. See, for example,

Salwen, H., and C. E. Grosch, *J. Fluid Mech.*, *54*, 93 (1972).
Gary, V. K., and W. T. Rouleau, *J. Fluid Mech.*, *54*, 113 (1972).

A number of flows of interest lead to variants of the Orr-Sommerfeld equation in which terms containing the parabolic profile $1 - y^2$ and its derivatives are replaced by other velocity fields and/or the boundary conditions are changed to reflect the existence of a free surface. The stability of plane Couette flow, which appears always to be stable to infinitesimal disturbances, is treated in

Gallagher, A. P., and A. M. Mercer, *J. Fluid Mech.*, *13*, 91 (1962); *18*, 350 (1964).

The stability of a thin liquid film on an inclined plane has been reviewed in detail in

Krantz, W. B., and S. L. Goren, *Ind. Eng. Chem. Fundamentals*, *10*, 91 (1971).

See also

Anshus, B. E., *Ind. Eng. Chem. Fundamentals*, *11*, 502 (1972).
Anshus, B. E., and E. Ruckenstein, to be published.
Bankoff, S. G., *Intern. J. Heat Mass Transfer*, *14*, 337 (1971).
Cerro, R. L., and S. Whitaker, *Chem. Eng. Sci.*, *26*, 742, 785 (1971).
Ladikov, Yu. N., *Fluid Mech.*, *5*, 465 (1973).

A perturbation analysis leads to an accurate analytical solution for the falling film problem. The extension to slightly viscoelastic liquids has been published a number of times:

Listrov, A. T., *J. Appl. Mech. Tech. Phys.*, *5*, 102 (1965).
Gupta, A. S., *J. Fluid Mech.*, *28*, 17 (1967).
Gupta, A. S., and L. Rai, *Proc. Cambridge Phil. Soc.*, *63*, 527 (1967).
Lai, W., *Phys. Fluids*, *10*, 844 (1967).

The analyses are all equivalent. Gupta's paper reaches some physically impossible conclusions and should be read in conjunciton with either Porteous and Denn, cited above, or

Craik, A. D. D., *J. Fluid Mech.*, *33*, 33 (1968).

The effect of three-dimensional disturbances on the falling film has been studied in

Gupta, A. S., and L. Rai, *J. Fluid Mech.*, *33*, 87 (1968).

For physically meaningful values of the parameters the two-dimensional disturbance is the more unstable.

The stability of a laminar boundary layer is discussed in Lin, Betchov and Criminale, and Radbill and McCue, cited above, and in

Schlichting, H., *Boundary Layer Theory*, 6th ed., McGraw-Hill, New York, 1968.

Ross, J. A., F. H. Barnes, J. G. Burns, and M. A. S. Ross, *J. Fluid Mech.*, *43*, 819 (1970).

Bouthier, M., *J. Mécanique*, *12*, 75 (1973).

See also a recent review paper on boundary layers,

Smith, A. M. O., *Appl. Mech. Rev.*, *23*, 1 (1970).

The study of the stability of stratified flow of two liquids leads to an Orr-Sommerfeld equation for each layer, with a coupling through the boundary condition at the interface. This is known as *Kelvin-Helmholtz instability*, for which an analytical solution can be found. See

Chandrasekhar, S., *Hydrodynamic and Hydromagnetic Stability*, Oxford University Press, Inc., New York, 1961.

Sontowski, J. F., B. S. Seidel, and W. F. Ames, *Quart. Appl. Math.*, *27*, 335 (1969).

For stratified gas-liquid flow in a pipe under some conditions of processing interest the appropriate assumptions reduce the problem to one equivalent to the falling film, and agreement with experiment is good. See

Russell, T. W. F., and A. W. Etchells, to be published; A. W. Etchells, Ph.D. Thesis, University of Delaware, Newark, 1971.

Finally, we note that there is considerable interest in the stability problem associated with the breakup of a liquid jet. The problem is summarized in Chandrasekhar, cited above, and comprehensively studied and reviewed in

Grant, R. P., and S. Middleman, *AIChE J.*, *12*, 669 (1966).

A more recent, but less accessible review, is in a Dutch Ph.D thesis, published in a limited edition as

Van de Sande, E., *Air Entrainment by Plunging Water Jets*, Delft University Press, Delft, Netherlands, 1974.

For a study of jets of viscoelastic liquids, see

Goldin, M., J. Yerushalmi, R. Pfeffer, and R. Shinnar, *J. Fluid Mech.*, *38*, 689 (1969).

Middleman, S., *Chem. Eng. Sci.*, *20*, 1037 (1965).

The effect of an electric field on jet stability is treated by

Hermans, J. J., *J. Electroanal. Chem.*, *37*, 337 (1972).

Saville, D. A., *Phys. Fluids*, *13*, 2987 (1970).

Linear Stability: Polymer Processing

12

12.1 Introduction

This chapter and the two which follow briefly consider some further applications of the linear stability analysis. No new principles are introduced, but the examples are of considerable engineering significance. The first two of these are extrusion and spinning of a molten polymer.

In melt extrusion the liquid is forced through a contracting entry into a die which may or may not be long enough for a fully developed laminar profile to form. For most applications a smooth surface is required on the extrudate which emerges from the die. It is found in practice that there is a limiting flow rate beyond which a smooth extrudate cannot be obtained. This phenomenon, often referred to in the literature as *melt fracture*, is a consequence of an instability of the steady laminar flow in the entry and the die. It is not clear whether it is the flow in the entry or in the die which first becomes unstable, and there is some evidence that this depends on the polymer. The onset appears to occur first in the inlet for branched polyethylene, for example, and in the die for linear polyethylene.

In melt spinning the extruded thread or film is stretched by a take-up device which is moving at a higher velocity than the material leaving the die. A limiting factor in spinning is an inability to maintain a time-invariant

thread diameter when the take-up velocity is too high. In controlled laboratory experiments this problem is manifested by a phenomenon known as
draw resonance, a smooth periodic fluctuation in fiber diameter and tension.
Diameter fluctuations of 30% are not uncommon.

The simplest reasonable description of the stress-deformation rate relation
in a polymer melt is a generalization of the Maxwell equation of linear
viscoelasticity known as the White-Metzner equation,

$$\tilde{\tau} + \frac{\mu}{G}\left[\frac{\partial \tilde{\tau}}{\partial \tilde{t}} + \tilde{\mathbf{v}} \cdot \tilde{\nabla}\tilde{\tau} - (\tilde{\nabla}\tilde{\mathbf{v}} \cdot \tilde{\tau}) - \tilde{\tau} \cdot (\tilde{\nabla}\tilde{\mathbf{v}})^T\right] = \mu[\tilde{\nabla}\tilde{\mathbf{v}} + (\tilde{\nabla}\tilde{\mathbf{v}})^T] \qquad (12\text{-}1)$$

μ is the viscosity and G the elastic shear modulus. Experimentally, the viscosity is a function of shear rate, but the shear modulus is often nearly constant.
In our analyses here we wish to determine the role played by fluid elasticity,
so we shall neglect the complication of shear dependency and take μ as a
constant.

12.2 Extrusion Instability

In looking for the mechanism of the melt extrusion instability we shall assume
that the die is sufficiently long to establish fully developed flow. We shall
consider the case of a cylindrical die of radius R with the axis in the horizontal
z direction and a constant pressure gradient dp/dz. Then the laminar flow,
which is a solution of Eq. (12-1) in conjunction with the Cauchy momentum
and continuity equations, (10-1) and (10-2), is

$$\tilde{v}_r = \tilde{v}_\theta = 0, \qquad \tilde{v}_z = 2V\left[1 - \left(\frac{\tilde{r}}{R}\right)^2\right]$$

$$\tilde{\tau}_{rr} = \tilde{\tau}_{zz} = \tilde{\tau}_{rz} = \tilde{\tau}_{\theta z} = 0$$

$$\tilde{\tau}_{r\theta} = \frac{-4\mu V r}{R^2}$$

$$\tilde{\tau}_{\theta\theta} = \frac{32\mu^2 V^2 r^2}{G R^4} \qquad\qquad (12\text{-}2)$$

$$\tilde{p} = \tilde{p}_0 + \frac{d\tilde{p}}{d\tilde{z}}\tilde{z}$$

V is the mean velocity. Experimental data for the melt instability seem to
correlate with a critical value of the elastic shear strain, S_R, defined as the
ratio of the wall shearing stress to the elastic shear modulus. (This ratio is

sometimes called the recoverable shear.) For the fluid model used in this study the elastic shear strain is

$$S_R = \frac{4\mu V}{RG} \tag{12-3}$$

The stability analysis proceeds in what should now be a familiar pattern. Equations (10-1), (10-2), and (12-1) are linearized about the steady solution, Eq. (12-2). We then seek a separation-of-variables solution to the linear partial differential equations in which the time dependence is of the form exp $(ikct)$. The details are tedious but straightforward, and we shall omit them here. Two assumptions are made. First, we assume that the disturbance is two-dimensional with no θ dependence. Second, because the Reynolds number is so small (of order 10^{-3}) at the onset of this instability, while S_R is of order 1 to 10, we neglect all terms containing Re. The result is the following linear eigenvalue problem, where $y = \tilde{r}/R$, the independent variable, is the dimensionless radial coordinate:

$$\psi^{IV} + \beta_3 \psi''' + \beta_2 \psi'' + \beta_1 \psi' + \beta_0 \psi = 0 \tag{12-4}$$

$$\psi(1) = \psi'(1) = 0$$

$$\lim_{y \to 0} \frac{\psi(y)}{y} = 0, \qquad \lim_{y \to 0} \frac{\psi'(y)}{y} = \text{finite} \tag{12-5}$$

$$\beta_3 = -\frac{2}{y} + \frac{2k^2 S_R^2 y(1 - y^2 - c)}{2 + ikS_R(1 - y^2 - c)} \tag{12-6a}$$

$$\beta_2 = -2k^2 + \frac{3}{y^2} + \frac{2k^4 S_R^4 y^2(1 - y^2 - c)^2}{[2 + ikS_R(1 - y^2 - c)]^2} \tag{12-6b}$$

$$\beta_1 = \frac{2k^2}{y} - \frac{3}{y^3} + \frac{2k^4 S_R^2 y(1 - y^2 - c)}{[2 + ikS_R(1 - y^2 - c)]^2}$$
$$\times [S_R^2(1 + 3y^2 - c) - 2 - ikS_R(1 - y^2 - c)] \tag{12-6c}$$

$$\beta_0 = k^4(1 + 2S_R^2 y^2) + \frac{2k^3 S_R^2}{[2 + ikS_R(1 - y^2 - c)]^2}$$
$$\times [4ky^2 + 8kS_R^2 y^4 + 4kS_R^2 y^2(1 - y^2 - c)] \tag{12-6d}$$

While the problem is linear, the eigenvalue, c, enters in a highly nonlinear fashion. Essentially the same nonlinearities arise in the equation for transition of a dilute polymer solution discussed in Sec. 11.6.

The solution was obtained using Galerkin's method with the approximating series

$$\psi = \sum_{n=1}^{N} C_n(1 - y^2)^2 y^{n+1} \qquad (12\text{-}7)$$

Because of the nonlinearities in Eq. (12-6), the calculation of each eigenvalue is an iterative process, and convergence was extremely slow. As a result, the neutral curve was calculated directly by setting c_I to zero, fixing k, and searching systematically over S_R and c_R to determine points where the real and imaginary parts of the determinant vanish simultaneously. The eigenvalues were found to be widely spaced, and the instability is determined by the eigenvalue of smallest modulus.

The neutral curve is shown in Fig. 12.1 for $N = 12$, 14, and 15 terms. With $N > 15$ numerical error was too large to allow unambiguous determination of the location of the zeros of the determinant. It is clear that convergence has not been attained, but the location of the minimum is insensitive

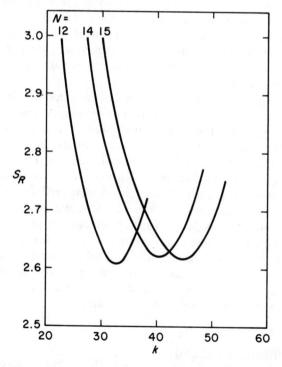

Figure 12.1 Curve of marginal stability for polymer extrusion with 12, 14, and 15 terms in Galerkin's method. [After R. Rothenberger, D. H. McCoy, and M. M. Denn, *Trans. Soc. Rheol.*, *17*, 259 (1973), with permission.]

to the number of terms. Thus, we can conclude that the flow must become unstable when S_R exceeds a critical value of approximately 2.6.

The next eigenvalue also has a neutral curve where c_I changes from positive to negative, but the minimum occurs at $S_R = 5.7$, so the first mode clearly governs the transition. Interestingly, the structure of the comparable problem between flat plates (a *slit die*) is somewhat different. There, the neutral curves for the seven eigenvalues which can be calculated all lie close to one another, and to within the accuracy of the calculation it is not possible to tell which is the critical one. The computed critical value of S_R in a slit die is 0.7.

Data on the onset of the melt flow instability scatter considerably, because of imprecision in methods of estimating the modulus, G, and the difficulty in precisely identifying the onset of the flow transition. All reported data can be interpreted in terms of a critical value of S_R, however, with the range $2 \leq S_{R_c} \leq 15$, and the most widely accepted value is $S_{R_c} \simeq 5$. This is reasonably good agreement with the theoretical value, particularly considering the fact that the theoretical analysis is based on the simplest possible description of the fluid stress distribution. There is some evidence in the literature that the critical elastic shear strain is a function of the molecular weight distribution, while the fluid model which we have used probably has its greatest validity for a monodisperse (single molecular weight) polymer system. A careful study by J. Vlachopoulos and M. Alam gives a critical value of $S_R = 2.65$ for a monodisperse system.

The assumption of a two-dimensional disturbance is a limiting one for this flow. A two-dimensional distortion is sometimes observed on the extrudate, but a spiraling, three-dimensional pattern is much more common. In addition, other mechanisms are possible to explain the onset of the instability. Some investigators believe that molten polymers can undergo slip at the wall, and it can be shown that under such conditions even a Newtonian fluid will become unstable at a critical flow rate. A more general stress constitutive relation also allows for a nonoscillatory Taylor vortex-like instability, and it has been shown that acceleration discontinuities can grow in a nonlinear viscoelastic liquid and cause a flow instability. These mechanisms may all be competitive, but the one reaching a critical value earliest will determine the transition in any given situation. Thus, the elastically induced transition computed here does represent a bound on the range of possible processing, though some other instability might occur earlier.

12.3 Spinning Instability

Several assumptions are made in describing fiber spinning. Inertia is neglected in the momentum balance, so the Cauchy momentum equation (10-1) simplifies to a requirement of constant tension along the thread, or

$$\frac{\partial}{\partial z}(aT) = 0 \tag{12-8}$$

a is the cross-sectional area of the thread, made dimensionless with respect to the initial area, and T is the tensile stress, made dimensionless with respect to the quantity $\mu \tilde{v}_f / L$. L is the length of the thread-line and $\tilde{v}_f$ the windup velocity. Distance along the thread, z, is made dimensionless with respect to L.

The tensile stress is assumed to be large compared to the stress normal to the stretching axis, and diagonal terms are assumed to dominate the deformation rate tensor, the right-hand side of Eq. (12-1). These two assumptions appear to be valid experimentally for highly elastic liquids. Then the stress constitutive equation (12-1) simplifies to

$$\frac{\partial T}{\partial t} + v \frac{\partial T}{\partial z} + \left(\frac{1}{\alpha} - 2\frac{\partial v}{\partial z}\right)T = \frac{2}{\alpha D_R}\frac{\partial v}{\partial z} \tag{12-9}$$

The velocity, v, is made dimensionless with respect to the initial velocity, $\tilde{v}_0$. The *drawdown ratio*, D_R, and the parameter including fluid elasticity, α, are defined as

$$D_R = \frac{\tilde{v}_f}{\tilde{v}_0} \qquad \alpha = \frac{\mu \tilde{v}_0}{LG} \tag{12-10}$$

Note the relation between α and the elastic shear strain, S_R, defined for extrusion by Eq. (12-3). Finally, the continuity equation can be written as

$$\frac{\partial a}{\partial t} + \frac{\partial av}{\partial z} = 0 \tag{12-11}$$

The boundary conditions used in most studies are

$$v(0) = a(0) = 1, \qquad v(1) = D_R \tag{12-12}$$

This corresponds to specifying the extrusion velocity and area and the windup velocity. The steady-state solution is implicit for any but a Newtonian fluid ($\alpha = 0$):

$$a_s v_s = 1 \tag{12-13}$$

$$a_s T_s = \frac{2 \ln D_R}{D_R [1 - \alpha(D_R - 1)]} \tag{12-14}$$

$$\frac{1 - \alpha(D_R - 1)}{\ln D_R} \ln v_s + \alpha(v_s - 1) = z \tag{12-15}$$

An interesting feature of the solution is that finite tensile stresses exist only for

$$D_R \leq 1 + \frac{1}{\alpha} \qquad (12\text{-}16)$$

Thus, according to the fluid model, for any value of α there is an upper bound on the attainable drawdown ratio, and for very large α the possible drawdown will be quite small.

Equations (12-8), (12-9), and (12-11) contain only quadratic nonlinearities, so linearization about the steady solution is straightforward. If we seek separation-of-variables solutions with a time dependence $\exp(\lambda t)$, we obtain the following linear ordinary differential equations:

$$\left[v_s + \frac{1 - \alpha(D_R - 1)}{\alpha \ln D_R}\right]\psi' = \left[\lambda - v_s' + \frac{1}{\alpha}\right]\tau + \lambda v_s^2 \phi \qquad (12\text{-}17\text{a})$$

$$\phi' = \frac{v_s'}{v_s^3}\tau - \frac{v_s'}{v_s}\phi - \frac{1}{v_s^2}\tau' \qquad (12\text{-}17\text{b})$$

$$\tau' = \lambda v_s \phi + \frac{v_s'}{v_s}\tau - \frac{v_s'}{v_s}\psi + \psi' \qquad (12\text{-}17\text{c})$$

$$\psi(0) = \psi(1) = \phi(0) = 0 \qquad (12\text{-}18)$$

ψ, ϕ, and τ refer to the deviations in velocity, area, and stress, respectively. Numerical values of $v_s(z)$ are obtained most simply by implicitly differentiating Eq. (12-15) to obtain the equation

$$v_s' = \frac{1}{\{[1 - \alpha(D_R - 1)]/(v_s \ln D_R)\} + \alpha} \qquad v_s(0) = 1 \qquad (12\text{-}19)$$

Equation (12-19) can then be numerically integrated to give v_s and v_s'. This is much easier than numerical solution of the algebraic equation (12-15).

This system of equations provides a nice example of an· alternative approach to numerical solution which, with generalization, has also been used on the problems studied in the two preceding chapters. The initial conditions on ψ and ϕ are zero. The initial condition on τ must be nonzero for a nontrivial solution. Because of the homogeneity of the differential equations, there is no loss of generality in taking $\tau(0) = 1$, and Eqs. (12-17) can then be integrated numerically. The problem is now reduced to a search over λ to find values such that $\psi(1) = 0$. If we focus first on finding the point of neutral stability, then we may set $\lambda_R = 0$ and carry out a one-dimensional search over λ_I. This is a very efficient procedure when the differential equations for the problem present no numerical difficulties, as is the case here. Points near the neutral point can then be computed to determine the sign of λ_R. It should be noted that ψ, ϕ and τ are all complex functions, and either complex arithmetic must be used in the numerical integrations or the real and imaginary parts of the equations must be separated. There is no loss in generality in taking the imaginary part of $\tau(0)$ as zero, since this simply determines the origin of the time scale.

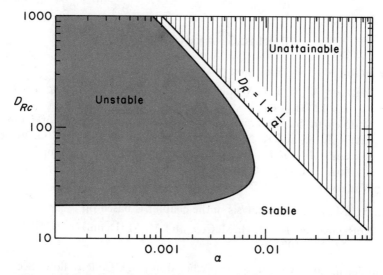

Figure 12.2 Regions of stable and unstable melt spinning in the drawdown ratio-viscoelastic parameter plane.

The critical drawdown ratio, D_{Rc}, is shown in Fig. 12.2 as a function of α. As $\alpha \to 0$ a critical value of $D_{Rc} \simeq 20$ is obtained. For $\alpha > 0.009$ the flow is stable to infinitesimal disturbances for any attainable drawdown ratio. The existence of a second stable region at large values of D_R is of considerable interest, since it suggests that extremely large drawdowns might be achieved if a means could be found to "spin through" the unstable region. The upper and lower branches of the D_{Rc}-α curve reflect grossly different kinematical regimes. Along the lower branch the first term on the left of Eq. (12-15) dominates, and the velocity is approximately exponential with distance, as in a Newtonian fluid. On the upper branch the second term dominates, and the velocity is nearly linear. The latter behavior is typical of experimental observations in highly elastic liquids.

This analysis is rather primitive, and extensions have been carried out in the inelastic limit ($\alpha = 0$) to account for effects of inertia, surface tension, variable viscosity, and heat transfer. There are little data in which all experimental parameters are sufficiently well defined for a meaningful comparison with the theory. A series of four experiments on molten polystyrene at several temperatures shows the onset of draw resonance in the range $3 \le D_R \le 5$, with $\alpha \simeq 0.05$–0.3. According to the theory, the flow should be stable, and it is not clear whether the discrepancy is because of a subcritical instability to finite disturbances or a sensitivity to the stress constitutive equation. The viscosity, μ, is highly shear-dependent in molten polystyrene, for example, while it is taken as constant in the analysis.

BIBLIOGRAPHICAL NOTES

For an introductory discussion of stress constitutive equations, see

> Metzner, A. B., J. L. White, and M. M. Denn, *Chem. Eng. Progr.*, *62*, no. 12, 81 (1966).
> Middleman, S., *The Flow of High Polymers*, Wiley, New York, 1968.
> Rivlin, R. S., and K. N. Sawyers, in M. Van Dyke and W. G. Vincenti, eds., *Annual Review of Fluid Mechanics*, vol. 3, Annual Reviews, Inc., Palo Alto, Calif., 1971.
> Walters, K., *Nature*, *207*, 826 (1965).

The extrusion stability analysis in the chapter is based on

> Rothenberger, R., D. H. McCoy, and M. M. Denn, *Trans. Soc. Rheol.*, *17*, 259 (1973).

A similar instability seems to be predicted in plane Couette flow. See

> Gorodstov, V. A., and A. I. Leonov, *J. Appl. Math. Mech.* (translation of *Prikl. Math. Mekh.*), *31*, 289 (1967).
> McIntire, L. V., *J. Appl. Polymer. Sci.*, *16*, 2901 (1972).

Some recent papers which survey the area of instability in polymer melt extrusion are

> Boger, D. V., and H. L. Williams, *Polymer Eng. Sci.*, *12*, 309 (1972).
> Pearson, J. R. A., *Plastics Polymers*, *37*, 285 (1969).
> Vinogradov, G. V., N. I. Insarova, B. B. Boiko, and E. K. Borisenkova, *Polymer Eng. Sci.*, *12*, 323 (1972).

The data mentioned in the text are in

> Vlachopoulos, J., and M. Alam, *Polymer Eng. Sci.*, *12*, 184 (1972).

Boger and Williams have shown that these data can be interpreted as leading to a critical value $S_R = 2.65$ in the limit of constant viscosity, which is one of the assumptions of the analysis.

The other mechanisms which it has been shown could lead to melt flow instability are slip at the wall, a Taylor vortex-like cellular instability, and growth of acceleration discontinuities. These are treated, respectively, by Pearson and Petrie, Bhatnagar and Giesekus, and Coleman and Gurtin, Dunwoody, and Sadd. It is doubtful that any of these is associated with melt fracture.

Pearson, J. R. A., and C. J. S. Petrie, in E. H. Lee, ed., *Proceedings of the 4th International Congress on Rheology*, part 3, Wiley-Interscience, New York, 1965.

Pearson, J. R. A., and C. J. S. Petrie, in R. W. Whorlow, ed., *Polymer Systems: Deformation and Flow*, Macmillan, New York, 1967.

Bhatnagar, R. K., and H. Giesekus, *Rheol. Acta*, *9*, 53, 412 (1970).

Coleman, B. D., and M. Gurtin, *J. Fluid Mech.*, *33*, 33 (1968).

Dunwoody, J., *Arch. Rat. Mech. Anal.*, *38*, 372 (1970).

Sadd, M. H., *Trans. Soc. Rheol.*, *17*, 647 (1973).

There is an overview of research on spinning of threads and sheets in *Transactions of the Society of Rheology*, *16*, no. 3 (1972). The analysis of the spinning instability and the data cited in the text are in

Zeichner, G. R., "Spinability of Viscoelastic Fluids," M.Ch.E. Thesis, University of Delaware, Newark, 1973.

The assumption that the tensile stress is large compared to the stress normal to the stretching axis is valid only in the limit of very large stress. A solution to the steady state spinning equations without this assumption has been obtained in an unpublished paper by Denn and Avenas, but the complete stability problem has not yet been treated. All the theoretical work for the inelastic liquid has been done by Pearson and coworkers and is summarized in

Pearson, J. R. A., and Y. T. Shah, *Trans. Soc. Rheol.*, *16*, 519 (1972).

See also

Han, C. D., R. R. Lamonte, and Y. T. Shah, *J. Appl. Polymer Sci.*, *16*, 3307 (1972).

Both the latter papers give references to earlier work and discuss additional experiments. An approach to the problem which is quite different from the traditional stability analyses described in the text is

Chang, H., and A. S. Lodge, *Rheol. Acta*, *10*, 448 (1971).

Some polymer processing instabilities may be related to an interaction between the wall heat transfer and the flow through a temperature-dependent viscosity. References to some relevant analyses are given in the Postface.

Linear Stability: 13
Convective Transport

13.1 Introduction

In Sec. 1.4 we described an experiment which shows that a temperature gradient can induce motion in a quiescent liquid and alter the mode of heat transfer from conduction to convection. We can now explain this phenomenon using linear stability theory. As we noted in Chapter 1, it is the variation in density which is the ultimate cause of the instability. Thus, fluid compressibility must be taken into account. The relevant physical property is the *coefficient of thermal expansion,*

$$\alpha = - \frac{d \ln \rho}{dT}$$

α is typically in the range 10^{-3}–10^{-4} reciprocal degrees centigrade. For simplicity we shall take α as a constant. We shall also take the viscosity and thermal conductivity as constants.

The geometry is shown in Fig. 1.7. The fluid is contained between two plates separated by a spacing δ. The lower plate is maintained at temperature

T_0, and the upper plate at a lower temperature $T_0 - \Delta$. The gravitational acceleration points directly downward in the negative y direction. If we choose the characteristic velocity as $\mu/\rho_0\delta$ and the characteristic time as $\delta^2\rho_0/\mu$, where ρ_0 is the density at $y = 0$, and make density and temperature dimensionless with respect to ρ_0 and Δ, respectively, then the continuity, momentum (Navier-Stokes), and energy equations for a compressible liquid are, respectively,

$$\frac{\partial \rho}{\partial t} + \mathbf{v}\cdot\nabla\rho = -\rho\nabla\cdot\mathbf{v} \tag{13-1}$$

$$\rho\frac{\partial \mathbf{v}}{\partial t} + \rho\mathbf{v}\cdot\nabla\mathbf{v} = -\nabla p + \nabla^2\mathbf{v} - \left(\frac{\rho_0^2\,\delta^3 g}{\mu^2}\right)\rho\mathbf{j} \tag{13-2}$$

$$\rho\frac{\partial T}{\partial t} + \rho\mathbf{v}\cdot\nabla T = \frac{1}{\mathrm{Pr}}\nabla^2 T \tag{13-3}$$

ρ, $\mathbf{v}$, and p are, respectively, dimensionless density, velocity, and pressure. T is the dimensionless temperature difference from T_0. g is the gravitational acceleration and κ the thermal diffusivity. $\mathbf{j}$ is the unit vector in the y direction. Pr is the Prandtl number,

$$\mathrm{Pr} = \frac{\mu}{\rho_0\kappa}$$

Viscous dissipation is neglected in the energy equation.

The steady-state solution to Eqs. (13-1) to (13-3) with a quiescent liquid and fixed temperatures at the boundaries is

$$\mathbf{v}_s = \mathbf{0} \tag{13-4a}$$

$$T_s = -y \tag{13-4b}$$

$$\rho_s = 1 + \alpha\Delta y \tag{13-4c}$$

$$p_s = p_0 - \left(\frac{\rho_0^2\,\delta^3 g}{\mu^2}\right)\left(y + \frac{\alpha\Delta}{2}y^2\right) \tag{13-4d}$$

The temperature decreases linearly between the plates, while the density increases linearly.

13.2 Stability Analysis

The perturbation variables are defined as follows:

$$\mathbf{u} = \mathbf{v} - \mathbf{v}_s = \mathbf{v}$$

$$\theta = T - T_s = T + y$$

$$\zeta = \rho - \rho_s = \rho - 1 - \alpha\Delta y$$

$$\eta = p - p_s = p - p_0 + \left(\frac{\rho_0^2 \delta^3 g}{\mu^2}\right)\left(y + \frac{\alpha\Delta}{2} y^2\right)$$

Substituting into Eqs. (13-1) to (13-3) and neglecting nonlinear terms in the perturbation variables lead to the linearized equations

$$\alpha\Delta \frac{\partial \theta}{\partial t} - \frac{\alpha\Delta}{\delta} w = (1 + \alpha\Delta y)\nabla \cdot \mathbf{u} \qquad (13\text{-}5)$$

$$\frac{\partial \mathbf{u}}{\partial t} = -\nabla\eta + \left(\frac{\mathrm{Ra}}{\mathrm{Pr}}\right) \theta\mathbf{j} + \nabla^2\mathbf{u} \qquad (13\text{-}6)$$

$$\frac{\partial \theta}{\partial t} = v + \frac{1}{\mathrm{Pr}} \nabla^2\theta \qquad (13\text{-}7)$$

u, v, and w are the x, y, and z components of $\mathbf{u}$, respectively. The Rayleigh number is defined as

$$\mathrm{Ra} = \frac{g\alpha\Delta\delta^3\rho_0}{\mu\kappa}$$

The term $\partial\theta/\partial t$ in the continuity equation (13-5) comes frcm the density-temperature relation

$$\frac{\partial \zeta}{\partial t} = \frac{d\rho}{dT} \frac{\partial \theta}{\partial t} = \alpha\Delta \frac{\partial \theta}{\partial t}$$

Boundary conditions are

$$u = v = w = 0, \qquad y = 0, 1 \qquad (13\text{-}8\mathrm{a})$$

$$\theta = 0, \qquad y = 0, 1 \qquad (13\text{-}8\mathrm{b})$$

together with boundedness requirements in the x and z directions.

One simplification is useful. For Δ of order several degrees, $\alpha\Delta$ will be negligible compared to unity. Thus, the $\alpha\Delta y$ term in Eq. (13-5) can be neglected. Moreover, since w/δ is of order $\nabla\cdot\mathbf{u}$, the term $\alpha\Delta w/\delta$ in that same equation can be neglected. Thus, the continuity equation becomes

$$\alpha\Delta\frac{\partial\theta}{\partial t} = \nabla\cdot\mathbf{u} \qquad (13\text{-}9)$$

This is known as the *Boussinesq approximation*. The approximation is equivalent to assuming that density variations are important only in the buoyancy term in the momentum equation and can be neglected otherwise. It is often made right at the start in natural convection analyses.

Solutions of Eqs. (13-6), (13-7), and (13-9) by separation of variables will be summations of terms of the form

$$u = v(y)e^{\lambda t}e^{ik_x x}e^{ik_z z}$$

$$v = \Phi(y)e^{\lambda t}e^{ik_x x}e^{ik_z z}$$

$$w = \omega(y)e^{\lambda t}e^{ik_x x}e^{ik_z z}$$

$$\theta = \psi(y)e^{\lambda t}e^{ik_x x}e^{ik_z z}$$

$$\eta = \beta(y)e^{\lambda t}e^{ik_x x}e^{ik_z z}$$

When these are substituted into the linear partial differential equations a set of linear ordinary differential equations is obtained. These can be combined, together with the following definitions,

$$k^2 = k_x^2 + k_z^2$$

$$\phi(y) = \text{Pr}\ \Phi(y)$$

into two equations,

$$\phi^{\text{IV}} - 2k^2\phi'' + k^4\phi - \lambda(\phi'' - k^2\phi) = \text{Ra}\ k^2\psi \qquad (13\text{-}10a)$$

$$\psi'' - k^2\psi - \lambda\ \text{Pr}\ \psi = -\phi \qquad (13\text{-}10b)$$

Boundary conditions are

$$\phi = \phi' = \psi = 0 \qquad \text{at } y = 0, 1 \qquad (13\text{-}11)$$

This system of equations is self-adjoint, and the eigenvalues, λ, can be shown to be real, as in Appendix D.3. Thus, the point of marginal stability is defined by $\lambda = 0$, and the equations for marginal stability reduce to

$$\phi^{\text{IV}} - 2k^2\phi'' + k^4\phi = \text{Ra}\ k^2\phi \qquad (13\text{-}12a)$$

$$\psi'' - k^2\psi = -\phi \qquad (13\text{-}12b)$$

Equations (13-11) and (13-12) are identical to Eqs. (10-30) and (10-31) for the stability of flow between rotating cylinders with $M = 0$, except that Ra replaces T as the critical parameter. According to the solution given by Eq. (10-34), there will be a transition at a critical Rayleigh number of $Ra_c = 1715$ in the first approximation, with a wave number $k_c = 3.12$. Retention of more terms adjusts the critical Rayleigh number slightly to 1708. This is in excellent agreement with the experimental data in Fig. 1.8, where the critical Rayleigh number is 1700 ± 50. The analysis predicts that the transition to convection should be independent of the Prandtl number, and this is also confirmed experimentally.

13.3 Concluding Remarks

The equation of mass conservation for a solute in a dilute binary system is identical to Eq. (13-3) with the Schmidt number replacing the Prandtl number, or, equivalently, mass instead of thermal diffusivity. Thus, identical arguments apply to convective instabilities induced by concentration gradients.

It is interesting to note that the analysis gives only a critical value of k but not of k_x and k_z. Thus, the theory does not predict the shape of the critical disturbance. There is evidence that the disturbance is two-dimensional in the form of roll cells, but the orientation of the cells may change at various locations in the fluid. When the upper surface is free (i.e., a liquid-air interface) hexagonal cells are observed.

The free surface problem was the first studied, following experiments in which H. Bénard observed hexagonal cells in a layer of spermaceti. A calculation identical to the one in this chapter but with the boundary condition of zero shear stress at the free surface gives a critical Rayleigh number of 1100 and a critical wave number of 2.68. With a free surface it is necessary to consider also surface tension effects, as is done in the next chapter. It is almost certain that Bénard's cells were not caused by a mechanism of the type discussed in this chapter. Nevertheless, the problem treated here is often called the Bénard problem. The first analysis of the problem, by Rayleigh, was stimulated by Bénard's experiments, and the name Bénard-Rayleigh problem is also used.

For a viscoelastic liquid a solution can be obtained in complete generality, because the flow perturbation is about a zero velocity, so all stress-deformation relations must reduce to conventional linear viscoelasticity. The parameter which enters is the complex viscosity, which is a measure of the stress response to small amplitude oscillations. For some values of the complex viscosity an oscillatory instability is possible, but the parameters seem to be beyond the range attained by real fluids. Theoretical and experimental studies of non-Newtonian inelastic liquids indicate that a shear thinning viscosity results in a lower critical Rayleigh number.

BIBLIOGRAPHICAL NOTES

The subject matter of this chapter is dealt with exhaustively in

Chandrasekhar, S., *Hydrodynamic and Hydromagnetic Stability*, Oxford University Press, Inc., New York, 1961.

Chandrasekhar's book contains theory, experiment, and history and should be consulted. Many of the fundamental papers are collected in a volume

Saltzman, B., *Selected Papers on the Theory of Thermal Convection*, Dover, New York, 1962.

This includes the original paper by Rayleigh,

Rayleigh, Lord, *Phil. Mag.*, *32*, 529 (1916).

The origins of the Boussinesq approximation are traced in a note in a paper by Joseph, who attributes the equations to A. Oberbeck.

Joseph, D. D., *J. Fluid Mech.*, *47*, 257 (1971).

A recent review with an extensive bibliography is

Whitehead, J. A., Jr., *Am. Scientist*, *59*, 444 (1971).

For the viscoelastic and non-Newtonian inelastic liquid analyses, see, respectively,

Sokolov, M., and R. I. Tanner, *Phys. Fluids*, *15*, 534 (1972).

Tien, C., H. S. Tsuei, and Z. S. Sun, *Intern. J. Heat Mass Transfer*, *12*, 1173 (1969).

Other extensions include heat generation (the earth's mantle), allowing for thermal diffusion (the *Soret effect*), radiation at the boundary, phase change, and flow in a porous medium (development of "salt fingers"). Recent papers for each, containing further references, are, respectively,

Turcotte, D. L., and E. R. Oxburgh, in M. Van Dyke and W. G. Vincenti, eds., *Annual Review of Fluid Mechanics*, vol. 4, Annual Reviews, Inc., Palo Alto, Calif., 1972.

Platten, J. K., and G. Chavepegar, *Phys. Fluids*, *15*, 1555 (1972).

Arpaci, V. S., and D. Gozum, *Phys. Fluids*, *16*, 581 (1973).

Busse, F. H., and G. Schubert, *J. Fluid Mech.*, *46*, 801 (1971).

Tauton, J. W., E. N. Lightfoot, and T. Green, *Phys. Fluids*, *15*, 748 (1972).

The effect of chemical reaction has been studied by Frisch and coworkers, with complete references given in

Bzdil, J., and H. Frisch, *Phys. Fluids*, *14*, 2048 (1971),

and by

Wankat, P. C., and W. R. Schowalter, *AIChE J.*, *17*, 1346 (1971); *18*, 769 (1972).

The latter paper also contains a review of oscillatory instability in convective flow.

Stability in heat transfer generally is covered in the annual reviews in *International Journal of Heat and Mass Transfer* by E. R. G. Eckert and co-workers. The stability of natural convection is reviewed in detail in survey papers by Gebhart,

> Gebhart, B., *Appl. Mech. Rev.*, *22*, 291 (1969); in T. F. Irvine, Jr., and J. P. Hartnett, eds., *Advances in Heat Transfer*, vol. 9, Academic Press, New York, 1973; and in M. Van Dyke and W. G. Vincenti, eds., *Annual Review of Fluid Mechanics*, vol. 5, Annual Reviews, Inc., Palo Alto, Calif., 1973.

The extension of the analysis in this chapter to containers of finite dimension has been carried out. For analytical results and references to earlier work and experiments, see the paper by Joseph cited above and

> Charlson, G. S., and R. L. Sani, *Intern. J. Heat Mass Transfer*, *13*, 1479 (1970); *14*, 2157 (1971).
> Jennings, P. A., and R. L. Sani, *J. Heat Transfer*, *94*, 234 (1972).
> Catton, I., *J. Heat Transfer*, *94*, 446 (1972).

The effect of a density gradient in a shearing flow has been studied by several workers and is summarized in

> McIntire, L. V., and W. R. Schowalter, *AIChE J.*, *18*, 102 (1972).

See also

> McIntire, L. V., *J. Appl. Poly. Sci.*, *16*, 2901 (1972).

For a Newtonian liquid the effects of shearing and buoyancy uncouple, but there is considerable interaction for a viscoelastic liquid.

Linear Stability: Surface Tension-Driven Transport

14

14.1 Introduction

In Sec. 1.5 we described a remarkable set of experiments in which the mass transfer coefficient depends on the direction in which transport is taking place. We noted at that time that the phenomenon could be explained by accounting for flows induced by surface tension variations. We shall not attempt to deal here with the complexities of the experimental situation of Olander and Reddy but with a much simpler physical configuration which still reveals the essential features.

The geometry to be considered is shown in Fig. 14.1. A layer of fluid of thickness δ contains a solute of concentration $\tilde{c}$. The solute moves through the free surface to the surroundings, perhaps by evaporation. The driving force for mass transfer is $\tilde{c}(\delta) - \tilde{c}_e$, where $\tilde{c}_e$ is the concentration in equilibrium with the surroundings. A steady state is maintained by replenishing the

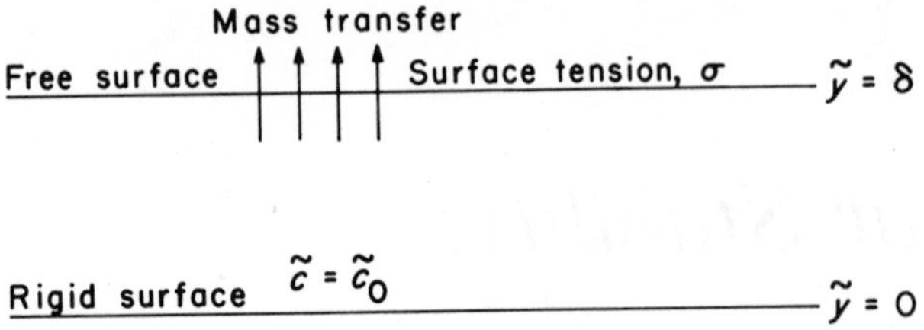

Figure 14.1. Schematic diagram for mass transfer through an interface.

lower surface to keep its concentration at $\tilde{c}_0$. The concentration distribution in the layer is found by solving the diffusion equation,

$$\frac{\partial \tilde{c}}{\partial \tilde{t}} + \tilde{\mathbf{v}} \cdot \tilde{\nabla} \tilde{c} = D \tilde{\nabla}^2 \tilde{c} \tag{14-1}$$

$$\tilde{c} = \tilde{c}_0 \qquad \text{at } \tilde{y} = 0 \tag{14-2a}$$

$$-D \frac{\partial \tilde{c}}{\partial \tilde{y}} = k_m (\tilde{c} - \tilde{c}_e) \qquad \text{at } \tilde{y} = \delta \tag{14-2b}$$

D is the mass diffusivity and k_m the mass transfer coefficient. The boundary condition (14-2b) is a mass balance at the free surface which is analogous to *Newton's law of cooling* for heat transfer. The steady-state solution to these equations is linear,

$$\tilde{c}_s(\tilde{y}) = \tilde{c}_0 - \frac{E(\tilde{c}_0 - \tilde{c}_e)}{1 + E} \frac{\tilde{y}}{\delta} \tag{14-3}$$

$$E = \frac{k_m \delta}{D}$$

E is sometimes known as the *Sherwood number*. In what follows it should be noted that the results carry over without change to a single-component system with temperature variations by simply replacing $\tilde{c}$, D, and k_m with temperature, thermal diffusivity, and the heat transfer coefficient, respectively.

It is convenient to define a dimensionless concentration as

$$c = \frac{1 + E}{E} \frac{\tilde{c} - \tilde{c}_0}{\tilde{c}_e - \tilde{c}_0}$$

Time, velocity, and pressure are made dimensionless as in the preceding chapter. The full set of transient mass and momentum conservation equations is then

$$\frac{\partial c}{\partial t} + \mathbf{v} \cdot \nabla c = \frac{1}{\text{Sc}} \nabla^2 c \tag{14-4}$$

$$\frac{\partial \mathbf{v}}{\partial t} + \mathbf{v} \cdot \nabla \mathbf{v} = -\nabla p + \nabla^2 \mathbf{v} \tag{14-5}$$

$$\nabla \cdot \mathbf{v} = 0 \tag{14-6}$$

Sc is the Schmidt number, $\mu/\rho D$, which is the mass transfer analog of the Prandtl number. To isolate the surface tension effect we have neglected buoyancy and taken the density as constant. Boundary conditions are

$$c = 0, \qquad \mathbf{v} = 0 \qquad \text{at } y = 0 \tag{14-7a}$$

$$\frac{\partial c}{\partial y} = 1 + E - Ec \qquad \text{at } y = 1 \tag{14-7b}$$

The mechanical boundary conditions at $y = 1$ are critical. One is no flow through the surface,

$$v = 0 \qquad \text{at } y = 1 \tag{14-7c}$$

where v is the y component of $\mathbf{v}$. The other is that the shear stresses at the surface are balanced by the surface tension gradients,

$$\tilde{\tau}_{xy} = \frac{\partial \sigma}{\partial \tilde{x}}, \qquad \tilde{\tau}_{xz} = \frac{\partial \sigma}{\partial \tilde{z}} \qquad \text{at } y = 1$$

In dimensional variables these can be expressed for a Newtonian fluid as

$$\mu \frac{\partial \tilde{u}}{\partial \tilde{y}} = \frac{\partial \sigma}{\partial \tilde{x}}, \qquad \mu \frac{\partial \tilde{w}}{\partial \tilde{y}} = \frac{\partial \sigma}{\partial \tilde{z}}$$

Differentiating the first with respect to x, the second with respect to z, and adding, we have

$$\frac{\partial^2 \sigma}{\partial \tilde{x}^2} + \frac{\partial^2 \sigma}{\partial \tilde{z}^2} = \mu \frac{\partial}{\partial \tilde{y}} \left(\frac{\partial \tilde{u}}{\partial \tilde{x}} + \frac{\partial \tilde{w}}{\partial \tilde{z}} \right) = -\mu \frac{\partial^2 \tilde{v}}{\partial \tilde{y}^2}$$

where the last equality comes from the continuity equation (14-6). The surface tension is a function of concentration. Define

$$S = -\frac{\partial \sigma}{\partial \tilde{c}}$$

Then, taking S as approximately constant, the surface boundary condition can be written as

$$\mu \frac{\partial^2 \tilde{v}}{\partial \tilde{y}^2} = S\left[\frac{\partial^2 \tilde{c}}{\partial \tilde{x}^2} + \frac{\partial^2 \tilde{c}}{\partial \tilde{y}^2}\right]$$

or, in dimensionless form,

$$\frac{\partial^2 v}{\partial y^2} = \frac{\text{Ma}}{\text{Sc}}\left[\frac{\partial^2 c}{\partial x^2} + \frac{\partial^2 c}{\partial z^2}\right] \qquad \text{at } y = 1 \qquad (14\text{-}7\text{d})$$

Ma is often called the *Marangoni* or *Thompson* number,

$$\text{Ma} = \frac{S\delta E(\tilde{c}_e - \tilde{c}_0)}{\mu D(1 + E)}$$

The steady-state solution to these equations is

$$c_s = y, \qquad \mathbf{v}_s = \mathbf{0}, \qquad p_s \quad = \text{constant} \qquad (14\text{-}8)$$

14.2 Stability Analysis

The perturbation variables are defined as

$$\mathbf{u} = \mathbf{v} - \mathbf{v}_s = \mathbf{v}$$
$$\xi = c - c_s = c + y$$
$$\eta = p - p_s = p - \text{constant}$$

The linearized analysis is then identical to that in the preceding chapter, leading finally to the differential equations

$$\phi^{\text{IV}} - 2k^2\phi'' + k^4\phi - \lambda(\phi'' - k^2\phi) = 0 \qquad (14\text{-}9\text{a})$$
$$\psi'' - k^2\psi - \lambda \, \text{Sc} \, \psi = -\phi \qquad (14\text{-}9\text{b})$$

Except for the buoyancy term, which we neglected, these are the same as Eqs. (13-10). The difference is in the boundary conditions, which become

$$\phi = \phi' = 0 \qquad \text{at } y = 0 \qquad (14\text{-}10\text{a})$$

$$\psi = 0 \qquad \text{at } y = 0 \qquad (14\text{-}10\text{b})$$

$$\phi = 0 \qquad \text{at } y = 1 \qquad (14\text{-}10\text{c})$$

$$\psi' + E\psi = 0 \qquad \text{at } y = 1 \qquad (14\text{-}10\text{d})$$

$$\phi'' + k^2 \, \text{Ma} \, \psi = 0 \qquad \text{at } y = 1 \qquad (14\text{-}10\text{e})$$

The principle of exchange of stabilities, $\lambda = 0$ at the transition, cannot be proved analytically for this system, but it can be established by numerical experimentation. If we assume that $\lambda = 0$, then an analytical solution for the critical Marangoni number can be obtained. The solution to Eq. (14-9a) which satisfies boundary conditions (14-10a) and (14-10c) is

$$\phi(y) = \sinh ky + \frac{k \cosh k - \sinh k}{\sinh k} y \sinh ky - ky \cosh ky$$

The solution to Eq. (14-9b) which satisfies boundary conditions (14-10b) and (14-10d) is then

$$\psi(y) = - \left\{ \frac{3}{4k} y \cosh ky + \frac{k \cosh k - \sinh k}{4k \sinh k} y^2 \cosh ky \right.$$

$$- \frac{1}{4} y^2 \sinh ky - \frac{k \cosh k - \sinh k}{4k^2 \sinh k} y \sinh ky$$

$$\left. - \sinh ky \left[\frac{k^2 \cosh^2 k + k \sinh k \cosh k + \sinh^2 k}{4k^2 \sinh k (k \cosh k + E \sinh k)} \right] \right\}$$

Substituting these two solutions into the remaining boundary condition (14-10e) leads finally to the relation

$$\text{Ma} = \frac{8k(k \cosh k + E \sinh k)(k - \sinh k \cosh k)}{k^3 \cosh k - \sinh^3 k} \qquad (14\text{-}11)$$

Equation (14-11) is plotted in Fig. 14.2 for several values of the Sherwood number. For a given system with fixed mass transfer characteristics there is a critical value of the Marangoni number beyond which the quiescent state is not stable. Above the critical value there will be a convective motion which

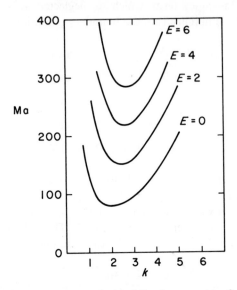

Figure 14.2. Curves of marginal stability for mass transfer through a surface for various values of the Sherwood number. [After J. R. A. Pearson, *J. Fluid Mech.*, *4*, 489 (1958), with permission of Cambridge University Press.]

will tend to increase the average surface concentration of solute and hence increase the driving force for mass transfer. The resulting increased rate of mass transfer will show up as an increase in the macroscopic overall mass transfer coefficient.

Quantitative comparison with experiment is difficult in this situation. We have assumed that the surface remains planar and have neglected the effect of the resistance of the surface to bending. This surface elasticity is very important, as is the influence of the Gibbs adsorption layer. In addition, when the transfer takes place between two liquids the instability depends in a complex way on the relative phase viscosities and diffusivities. An oscillatory instability is possible under some circumstances.

Clearly, it is possible to combine the analysis of this and the preceding chapter and treat buoyancy and surface tension-driven flows together. If Ra_c and Ma_c represent the critical Rayleigh and Marangoni numbers, respectively, in the absence of the other effect, then for $E = 0$ the critical condition taking both effects into account is approximately

$$\frac{Ra}{Ra_c} + \frac{Ma}{Ma_c} = 1$$

The region where each effect will be important can be discriminated by noting that Ma varies with δ and Ra with δ^3. Thus, in thin liquid films with free surfaces the surface tension mechanism will tend to dominate, while the buoyancy mechanism will dominate in thick films. Pearson estimates, based on the properties of typical liquids, that the effects become comparable in a liquid layer of about 1 cm.

BIBLIOGRAPHICAL NOTES

The first systematic experimental study of the surface tension-driven instability seems to be

Block, M. J., *Nature, 178,* 650 (1958).

The problem was first studied analytically by Pearson, and the chapter follows his paper,

Pearson, J. R. A., *J. Fluid Mech., 4,* 489 (1958).

An independent analysis considering the effect of two liquid phases is

Sternling, C. V., and L. E. Scriven, *AIChE J., 5,* 514 (1959).

For a broad review, see

Berg, J. C., A. Acrivos, and M. Boudart, *Adv. Chem. Eng., 6,* 61 (1966).

The calculations establishing the validity of the principle of exchange of stabilities are in

Vidal, A., and A. Acrivos, *Phys. Fluids., 9,* 615 (1966).

Generalizations of the analysis here accounting for combined buoyancy and surface tension effects, surface deformation, surface elasticity from surface active agents, Gibbs adsorption, chemical reaction, gravity waves, and viscoelasticity are, respectively,

Nield, D. A., *J. Fluid Mech., 19,* 341 (1964).
Scriven, L. E., and C. V. Sternling, *J. Fluid Mech., 19,* 321 (1964).
Berg, J. C., and A. Acrivos, *Chem. Eng. Sci., 20,* 737 (1965).
Brian, P. L. T., *AIChE J., 17,* 765 (1971); *18,* 231, 582 (1972).
Ruckenstein, E., and C. Berbente, *Chem. Eng. Sci., 19,* 329 (1964).
Smith, K. A., *J. Fluid Mech., 24,* 401 (1966).
Jarrell, M. S., University of Delaware, Newark, term paper to be published.

Scriven and Sternling point out the difference in cellular flow for the two instability mechanisms. Jarrell's analysis is completely general, with calculations for a Maxwell fluid. Overstability cannot occur in this case, and the

Newtonian result applies exactly. The combined effects of density and surface tension are studied experimentally in

> Heines, H., and J. W. Westwater, *Intern. J. Heat Mass Transfer*, *15*, 2109 (1972).
> Palmer, H. J., and J. C. Berg, *J. Fluid Mech.*, *47*, 779 (1971).

Heines and Westwater also review the general area. The combined buoyancy and surface tension analysis for two liquid phases, together with a review of the area, is in

> Zeren, R. W., and W. C. Reynolds, *J. Fluid Mech.*, *53*, 305 (1972).

Liapunov's
Direct Method

15

15.1 Introduction

Our concerns in dealing with distributed parameter systems are much the same as for lumped systems. Linear stability theory identifies conditions under which any infinitesimal disturbance must grow. Hence, linear theory defines a strict upper bound on the stability region, since it is impossible to maintain an equilibrium which is unstable to an arbitrarily small perturbation. Linear stability theory gives no information, however, about the effect of finite disturbances.

The procedures developed in Chapters 5 and 6 for determining the behavior of systems subject to finite-amplitude perturbations carry over directly to distributed parameter processes. In this chapter we shall illustrate the use of Liapunov's direct method with the catalytic reaction and rotational Couette flow. Only quadratic Liapunov functions seem to have been widely used in applications of the method to distributed systems. In part this is probably because in fluid mechanical problems the kinetic energy of the disturbance is a sum-of-squares of velocity deviations and hence is a natural Liapunov function with an identifiable physical meaning. In the fluid mechanics literature Liapunov's direct method is often called the *energy method* and can be traced to an independent development by W. M. Orr.

15.2 Catalytic Reaction

The equations for a catalytic reaction are developed in Chapter 9. For simplicity we shall restrict ourselves here to Lewis number unity and adiabatic perturbations. In that case Eq. (9-10) for the temperature can be written as

$$\frac{\partial \eta}{\partial t} = \frac{\partial^2 \eta}{\partial z^2} + \alpha \phi(y_s, \eta)\eta \tag{15-1}$$

Here, $y_s(z)$ is the steady-state temperature profile and η the temperature deviation. The function ϕ is defined as

$$\phi(y_s, \eta) = \frac{F(y_s + \eta) - F(y_s)}{\eta} \tag{15-2}$$

where $F(y)$ is defined by Eq. (9-11). In the limit of infinitesimal disturbances we have

$$\lim_{\eta \to 0} \phi(y_s, \eta) = F'(y_s(z)) \tag{15-3}$$

Boundary conditions for Eq. (15-1) are

$$\frac{\partial \eta}{\partial z} = 0 \quad \text{at } z = 0, \quad \eta = 0 \quad \text{at } z = 1 \tag{15-4}$$

As a Liapunov function we take the *spatial integral* of the square of the deviation,

$$V = \frac{1}{2} \int_0^1 \eta^2 \, dz \tag{15-5}$$

This is, of course, positive definite. The time derivative is

$$\dot{V} = \frac{1}{2} \frac{d}{dt} \int_0^1 \eta^2 \, dz = \frac{1}{2} \int_0^1 \frac{\partial \eta^2}{\partial t} \, dz = \int_0^1 \eta \frac{\partial \eta}{\partial t} \, dz$$

From Eq. (15-1) we can write

$$\dot{V} = \int_0^1 \left[\eta \frac{\partial^2 \eta}{\partial z^2} + \alpha \phi(y_s, \eta)\eta^2 \right] dz$$

If the first term is integrated by parts and the boundary conditions (15-4) are introduced, we obtain, finally,

$$\dot{V} = \int_0^1 \left[-\left(\frac{\partial \eta}{\partial z}\right)^2 + \alpha \phi(y_s, \eta)\eta^2 \right] dz \leq 0 \qquad (15\text{-}6)$$

Satisfaction of the inequality in Eq. (15-6) is the sufficient condition for stability.

It is possible to proceed with Eq. (15-6) without further approximation, and we shall do so in the next section. At this point, however, it is convenient to restrict attention to small disturbances and replace $\phi(y_s, \eta)$ with $F'(y_s)$. Using Poincaré's inequality (Appendix C) we can write

$$\int_0^1 \left(\frac{\partial \eta}{\partial z}\right)^2 dz \geq \frac{\pi^2}{4} \int_0^1 \eta^2 \, dz$$

in which case Eq. (15-6) becomes

$$\int_0^1 \left[-\left(\frac{\partial \eta}{\partial z}\right)^2 + \alpha F'(y_s)\eta^2 \right] dz \leq \int_0^1 \left[-\frac{\pi^2}{4} + \alpha F'(y_s) \right] \eta^2 \, dz \leq 0$$

This inequality is satisfied and the system is stable to small disturbances if

$$\alpha \max F'(y_s) \leq \frac{\pi^2}{4} \qquad (15\text{-}7)$$

which is the result obtained from the linear eigenvalue analysis in Sec. 9.2.

A somewhat different result is obtained by using Gavalas's inequality (Appendix C),

$$\eta^2 \leq (1 - z) \int_0^1 \left(\frac{\partial \eta}{\partial z}\right)^2 dz$$

If we let $z \in \Sigma$ denote the section of the interval $0 \leq z \leq 1$ for which $F'(y_s) > 0$ and $z \notin \Sigma$ the remainder of the interval, then the inequality can be combined with Eq. (15-6) as follows:

$$\dot{V} = \alpha \int_{z \notin \Sigma} F'(y_s)\eta^2 \, dz - \int_0^1 \left(\frac{\partial \eta}{\partial z}\right)^2 dz + \alpha \int_{z \in \Sigma} F'(y_s)\eta^2 \, dz$$

$$\leq \alpha \int_{z \notin \Sigma} F'(y_s)\eta^2 \, dz - \int_0^1 \left(\frac{\partial \eta}{\partial z}\right)^2 dz + \alpha \int_{z \in \Sigma} F'(y_s)(1 - z) \left[\int_0^1 \left(\frac{\partial \eta}{\partial z}\right)^2 dz \right] dz$$

$$= \alpha \int_{z \notin \Sigma} F'(y_s)\eta^2 \, dz - \left\{ 1 - \alpha \int_{z \in \Sigma} (1 - z)F'(y_s) \, dz \right\} \int_0^1 \left(\frac{\partial \eta}{\partial z}\right)^2 dz \leq 0$$

The inequality is satisfied and the system is stable to small disturbances if

$$\alpha \int_{z \in \Sigma} (1 - z)F'(y_s) \, dz \leq 1 \qquad (15\text{-}8)$$

This condition is not equivalent to Eq. (15-7), and it may be possible to prove stability using one approach when it is not possible using the other.

The approach to Liapunov stability taken here can be extended to the general case of $\mathscr{L} \neq 1$ and nonadiabatic perturbations. The results which have been obtained are quite conservative and of little practical value, however; and we shall not carry the analysis any further here.

15.3 Variational Approach

An estimate of the stability radius of the quadratic Liapunov function can be obtained by following the procedure outlined in Sec. 5.5. For simplicity we shall again limit ourselves to the adiabatic perturbation, $\mathscr{L} = 1$, but this restriction can be easily relaxed. The result is an estimate of a finite stability region which is calculated entirely from the solution to a linear, self-adjoint problem.

The starting point is Eq. (15-6), which is rewritten as

$$\frac{1}{\alpha} \geq \frac{\int_0^1 \phi(y_s, \eta)\eta^2 \, dz}{\int_0^1 (\partial\eta/\partial z)^2 \, dz}$$

A given steady state is guaranteed to be stable as long as α satisfies this inequality. The largest value of α for stability is therefore determined by the maximum of the right-hand side for functions satisfying Eq. (15-1) and the boundary conditions (15-4). A conservative estimate is obtained by widening the class of admissible functions to include all continuous differentiable functions which satisfy the boundary conditions. Thus, we have

$$\frac{1}{\alpha} \geq \frac{1}{\lambda} = \max \frac{\int_0^1 \phi(y_s, \eta)\eta^2 \, dz}{\int_0^1 (\partial\eta/\partial z)^2 \, dz} \qquad (15\text{-}9)$$

where the maximization is over all functions satisfying Eq. (15-4).

Determination of the maximum of the ratio of integrals in Eq. (15-9) is a problem in the calculus of variations. Using standard procedures of that theory it can be shown that the function which causes $1/\lambda$ to take on its maximum is a solution of the *Euler differential equation*

$$v'' + \lambda \phi(y_s, v)v + \lambda \frac{1}{2} \frac{\partial \phi(y_s, v)}{\partial v} v^2 = 0 \tag{15-10}$$

$$v'(0) = v(1) = 0 \tag{15-11}$$

This is a nonlinear eigenvalue problem, which always admits the trivial solution $v = 0$. The smallest positive eigenvalue λ for which a nontrivial solution exists is an upper bound on α for stability, while the eigenfunction $v(z)$ represents the destabilizing disturbance for given λ. If all eigenvalues are negative, then the integral $\int_0^1 \phi(y_s, \eta)\eta^2 \, dz$ is negative, the inequality (15-6) is always satisfied, and the steady state $y_s(z)$ is stable to disturbances of any magnitude.

In developing an approximate solution it is helpful to consider the linear approximation to Eq. (15-10),

$$v_0'' + \lambda_0 F'(y_s)v_0 = 0 \tag{15-12}$$

This corresponds to the point of marginal stability ($\lambda = 0$) in Eq. (9-19). Since the norm is arbitrary in a linear system, we shall take v_0 to have a unit norm,

$$\int_0^1 v_0^2(z) \, dz = 1 \tag{15-13}$$

If we take the solution to Eq. (15-10) as having norm $\mathscr{A}$,

$$\int_0^1 v^2(z) \, dz = \mathscr{A}^2 \tag{15-14}$$

then, following the outline in Appendix B.2, the eigenvalue λ and inequality (15-9) can be written to first order in $\mathscr{A}$ as

$$\alpha \leq \lambda = \lambda_0 \left[1 - \frac{3}{4} \frac{\int_0^1 F''(y_s)v_0^3 \, dz}{\int_0^1 F'(y_s)v_0^2 \, dz} \mathscr{A} + 0(\mathscr{A}^2) \right] \tag{15-15}$$

The result is directional, in that the eigenvalue depends on whether $v_0(0)$ is positive or negative. If we consider the amplitude to be positive, then the conservative bound is obtained by using the absolute value of the coefficient of $\mathscr{A}$. Equation (15-15) can then be rearranged to give the estimate of the amplitude, $\mathscr{A}$, for which stability is assured:

$$\mathscr{A} \le \frac{4}{3} \left| \frac{\int_0^1 F'(y_s) v_0^2(z)\, dz}{\int_0^1 F''(y_s) v_0^3(z)\, dz} \right| \left\{ 1 - \frac{\alpha}{\lambda_0} \right\} \tag{15-16}$$

This estimate of the radius of the Liapunov stability region is obtained entirely by solving a linear self-adjoint eigenvalue problem, Eq. (15-12) with boundary conditions (15-11) and normalization (15-13). The extension to the general case $\mathscr{L} \ne 1$ can be readily carried out, and it, too, requires solution only of a self-adjoint linear eigenvalue problem.

15.4 Rotational Couette Flow

We now turn to the stability of solutions of the Navier-Stokes equations to disturbances of finite amplitude. The general approach is like that of the two preceding sections. Because the nonlinearities in the Navier-Stokes equations are only quadratic, some analytical results can be obtained without approximation. On the other hand, the fact that we are dealing with four partial differential equations means that a good deal of manipulation is required.

The development here has been carried out by J. Serrin for a general class of flows, and we shall follow the general treatment for a while before specializing to the rotational Couette flow. The starting point is the Navier-Stokes and continuity equations for incompressible Newtonian fluid flow, Eqs. (10-4) and (10-2), respectively,

$$\rho \frac{\partial \mathbf{v}}{\partial t} + \rho \mathbf{v} \cdot \nabla \mathbf{v} = -\nabla p + \mu \nabla^2 \mathbf{v} + \rho \mathbf{f} \tag{15-17}$$

$$\nabla \cdot \mathbf{v} = 0 \tag{15-18}$$

We shall not bother with the tilde to denote dimensional quantities, since there is no advantage to making the equations dimensionless. The boundary conditions are no-slip on solid surfaces and boundedness in any direction of infinite extent (e.g., the axial direction between rotating cylinders). *We shall impose the further restriction that we shall consider only flows which are periodic in a direction of infinite extent.*

The steady-state solution to Eqs. (15-17) and (15-18) are denoted by $\mathbf{v}_s$ and p_s, and the perturbation variables $\mathbf{u}$ and y are defined by

$$\mathbf{u} = \mathbf{v} - \mathbf{v}_s$$

$$y = \frac{(p - p_s)}{\rho}$$

Because of the no-slip boundary condition, $\mathbf{u}$ vanishes identically on all solid surfaces. Substituting these definitions into the dynamic equations leads directly to an equivalent set of equations in terms of the perturbation variables,

$$\frac{\partial \mathbf{u}}{\partial t} + (\mathbf{v}_s + \mathbf{u}) \cdot \nabla \mathbf{u} + \mathbf{u} \cdot \nabla \mathbf{v}_s = -\nabla y + \nu \nabla^2 \mathbf{u} \tag{15-19}$$

$$\nabla \cdot \mathbf{u} = 0 \tag{15-20}$$

ν is the kinematic viscosity, μ/ρ. We emphasize again that $\mathbf{u}$ and y are assumed to be periodic in any direction of infinite extent.

As a Liapunov function we use the total kinetic energy of the disturbance,

$$V = \frac{1}{2} \int_{\mathscr{V}} \mathbf{u} \cdot \mathbf{u} \, d\mathscr{V} = \frac{1}{2} \int_{\mathscr{V}} u^2 \, d\mathscr{V} \tag{15-21}$$

u is the magnitude of the vector, $\mathbf{u}$. The region $\mathscr{V}$ is bounded by the solid surfaces and extends one wavelength in any direction of infinite extent. In taking the time derivative it is important to note that changes will occur both because of variations of $\mathbf{u}$ within $\mathscr{V}$ and because of flow across the boundary $\mathscr{S}$ of the region $\mathscr{V}$. This is expressed by means of the *Reynolds transport theorem* for differentiating the integral,

$$\dot{V} = \int_{\mathscr{V}} \mathbf{u} \frac{\partial \mathbf{u}}{\partial t} \, d\mathscr{V} + \frac{1}{2} \int_{\mathscr{S}} u^2 \mathbf{v} \cdot \mathbf{n} \, d\mathscr{S} \tag{15-22}$$

$\mathbf{n}$ is the outward normal along the boundary. The surface integral in this equation vanishes. $\mathbf{u}$ is zero along solid boundaries. $\mathbf{u}$ and $\mathbf{v}$ are periodic in directions of infinite extent, while $\mathbf{n}$ changes algebraic sign at opposite ends of a wavelength, so integrals along surfaces in directions of infinite extent cancel one another. Thus, substituting Eq. (15-19) into (15-22) the time derivative of the Liapunov function can be written as

$$\dot{V} = \int_{\mathscr{V}} \mathbf{u} \cdot [\nu \nabla^2 \mathbf{u} - \mathbf{v} \cdot \nabla \mathbf{u} - \mathbf{u} \cdot \nabla \mathbf{v}_s - \nabla y] \, d\mathscr{V} \tag{15-23}$$

Stability is ensured if we can determine conditions for which $\dot{V} \leq 0$.

At this point is it quite helpful to introduce several vector identities, as follows:

$$\mathbf{u} \cdot \nabla^2 \mathbf{u} = \nabla \cdot (\nabla \tfrac{1}{2} u^2) - \nabla \mathbf{u} : \nabla \mathbf{u}$$

$$\mathbf{u} \cdot (\mathbf{v} \cdot \nabla \mathbf{u}) = \nabla \cdot (\tfrac{1}{2} u^2 \mathbf{v}) - \tfrac{1}{2} u^2 \nabla \cdot \mathbf{v}$$

$$\mathbf{u} \cdot \nabla y = \nabla \cdot \mathbf{u} y - y \nabla \cdot \mathbf{u}$$

$$\mathbf{u} \cdot (\mathbf{u} \cdot \nabla \mathbf{v}_s) = \mathbf{u} \cdot \mathbf{D} \cdot \mathbf{u}$$

$$\mathbf{D} \equiv \tfrac{1}{2} [\nabla \mathbf{v}_s + (\nabla \mathbf{v}_s)^T]$$

$\mathbf{D}$ is the rate of strain tensor for the steady-state flow. The terms $\nabla \cdot \mathbf{u}$ and $\nabla \cdot \mathbf{v}$ are both equal to zero from Eqs. (15-18) and (15-20). All divergence terms can be grouped together,

$$\mathbf{\Phi} = \nabla \tfrac{1}{2} u^2 + \tfrac{1}{2} u^2 \mathbf{v} + \mathbf{u} y$$

and Eq. (15-23) can be rewritten as

$$\dot{V} = - \int_{\mathscr{V}} (\mathbf{u} \cdot \mathbf{D} \cdot \mathbf{u} + \nu \nabla \mathbf{u} : \nabla \mathbf{u}) \, d\mathscr{V} + \int_{\mathscr{V}} \nabla \cdot \mathbf{\Phi} \, d\mathscr{V}$$

We can use Green's theorem to write

$$\int_{\mathscr{V}} \nabla \cdot \mathbf{\Phi} \, d\mathscr{V} = \int_{\mathscr{S}} \mathbf{\Phi} \cdot \mathbf{n} \, d\mathscr{S}$$

and the surface integral vanishes for the reasons noted above. Thus, we obtain, finally, a relation known as the *Reynolds-Orr equation,*

$$\dot{V} = - \int_{\mathscr{V}} (\mathbf{u} \cdot \mathbf{D} \cdot \mathbf{u} + \nu \nabla \mathbf{u} : \nabla \mathbf{u}) \, d\mathscr{V} \qquad (15\text{-}24)$$

We now turn to the specific problem of rotational Couette flow. The geometry and basic flow is described in Sec. 10.1. The only nonzero component of $\mathbf{v}_s$ is the θ component, which is given by the relation

$$v_{\theta s} = Ar + \frac{B}{r}$$

The constants are defined in Eq. (10-11b). We note here that

$$B = \frac{(\Omega_1 - \Omega_2)R^2}{1 - [R/(R + \delta)]^2}$$

Then, taking the gradient in cylindrical coordinates,

$$\mathbf{D} = -\frac{B}{r^2} \begin{pmatrix} 0 & 1 & 0 \\ 1 & 0 & 0 \\ 0 & 0 & 0 \end{pmatrix}$$

$$\mathbf{u} \cdot \mathbf{D} \cdot \mathbf{u} = -\frac{2B}{r^2} u_1 u_2$$

and, from Eq. (15-24), a sufficient condition for stability is

$$\frac{2B}{v} \int_{\mathscr{V}} \frac{u_1 u_2}{r^2} \, d\mathscr{V} - \int_{\mathscr{V}} \nabla \mathbf{u} : \nabla \mathbf{u} \, d\mathscr{V} \leq 0 \qquad (15\text{-}25)$$

We shall now bound the terms in the inequality. For the first term we use the relation

$$(|u_1| - |u_2|)^2 = u_1^2 + u_2^2 - 2|u_1||u_2| \geq 0$$

This implies that

$$2Bu_1 u_2 \leq 2|B||u_1||u_2| \leq |B|(u_1^2 + u_2^2)$$
$$\leq |B|(u_1^2 + u_2^2 + u_3^2) = |B|u^2$$

Thus, inequality (15-25) is satisfied if

$$\frac{|B|}{v} \int_{\mathscr{V}} \frac{u^2}{r^2} \, d\mathscr{V} - \int_{\mathscr{V}} \nabla \mathbf{u} : \nabla \mathbf{u} \, d\mathscr{V} \leq 0 \qquad (15\text{-}26)$$

Serrin's inequality, proved in Appendix C.3, is

$$\int_{\mathscr{V}} \nabla \mathbf{u} : \nabla \mathbf{u} \, d\mathscr{V} \geq \left[\frac{\pi}{\ln \left((R + \delta)/R \right)} \right]^2 \int_{\mathscr{V}} \frac{u^2}{r} \, d\mathscr{V}$$

in which case inequality (15-26) is satisfied if

$$\left\{ \frac{|B|}{v} - \frac{\pi^2}{[\ln \left((R + \delta)/R \right)]^2} \right\} \int_{\mathscr{V}} \frac{u^2}{r} \, d\mathscr{V} \leq 0 \qquad (15\text{-}27)$$

The integral is positive, so a sufficient condition for stability is negativity of the coefficient. Using the definition of B the sufficient condition is

$$\left| \frac{\Omega_1 - \Omega_2}{v} \right| \leq \left[\left(\frac{R + \delta}{R} \right)^2 - 1 \right] \frac{\pi^2}{(R + \delta)^2 [\ln \left((R + \delta)/R \right)]^2} \qquad (15\text{-}28)$$

In the small gap approximation ($\delta \ll R$) used in the linear analysis this simplifies to

$$\frac{\delta}{R} \to 0 : \quad \left| \frac{\Omega_1 - \Omega_2}{v} \right| \leq \frac{2\pi^2}{R\delta} \qquad (15\text{-}29)$$

In terms of the Taylor number, defined by Eq. (10-28), and $M = 1 - \Omega_2/\Omega_1$ this criterion for stability can be rewritten as

$$\frac{\delta}{R} \to 0: \quad T \leq \frac{8\pi^4}{M}\frac{\delta}{R} \tag{15-30}$$

This is to be compared with the critical Taylor number from linear theory given in the first approximation by Eq. (10-34),

$$T_c = \frac{3430}{2 - M}$$

The result obtained from the Liapunov analysis is quite conservative. For $M = 1$ and $R/\delta = 40$, for example, the sufficient condition for stability is $T \leq 20$, while the transition occurs experimentally at the linear theory value of 3430. However, we have used very weak bounds in passing from the exact inequality (15-25) to the sufficient condition (15-28). A much stronger result can be obtained by using the variational approach developed in the preceding section. For positive B, Eq. (15-25) can be written as

$$\frac{v}{B} \geq \frac{\displaystyle\int_{\mathscr{V}} (2u_1 u_2/r^2)\, d\mathscr{V}}{\displaystyle\int_{\mathscr{V}} \nabla\mathbf{u}:\nabla\mathbf{u}\, d\mathscr{V}} \tag{15-31}$$

Then a sufficient condition for stability is $B/v \leq \lambda$, where

$$\frac{1}{\lambda} = \max \frac{\displaystyle\int_{\mathscr{V}} (2u_1 u_2/r^2)\, d\mathscr{V}}{\displaystyle\int_{\mathscr{V}} \nabla\mathbf{u}:\nabla\mathbf{u}\, d\mathscr{V}} \tag{15-32}$$

and the maximization is over all functions satisfying the boundary conditions and the continuity equation, $\nabla \cdot \mathbf{u} = 0$. This is analogous to Eq. (15-9) for the catalyst particle. The Euler equations for this quadratic problem in the calculus of variations are linear and are quite similar to the linearized stability equations in Chapter 10. The solution obtained by Serrin gives a sufficient condition for stability:

$$\left|\frac{\Omega_1 - \Omega_2}{v}\right| \leq 41.2\, \frac{2R + \delta}{\delta R(R + \delta)} \xrightarrow[\frac{\delta}{R} \to 0]{} \frac{82.4}{R\delta} \tag{15-33}$$

This is a 17.4-fold increase in the critical Taylor number in comparison to the simple result given by Eq. (15-29).

The variational Liapunov approach has been applied to a large number of flow problems by D. D. Joseph and coworkers. They have extended the analysis given here for the rotational Couette flow by using the integral of a weighted sum of squares of the components of $\mathbf{u}$ and have shown that over a wide range of values of M the result given in Sec. 10.4 using linear theory represents a sufficient condition for stability to finite disturbances. They have also established, for the convective transport problem treated by linear theory in Chapter 13, that the quiescent state is stable to finite disturbances below the critical Rayleigh number of 1708 obtained from linear theory.

We note, finally, that the Liapunov approach also provides the basis of a uniqueness proof for the steady state. A steady-state solution to the Navier-Stokes equations must be periodic in a direction of infinite extent, since the result must be independent of the arbitrary choice of an origin. Thus, if there is assumed to be another steady state, then the difference beween the two steady states can be taken as $\mathbf{u}$ in Eq. (15-19). But for a steady state, $\partial \mathbf{u}/\partial t = 0$ and $\dot{V} = 0$. A sufficient condition for uniqueness, then, is $\dot{V} < 0$ for all $\mathbf{u} \neq \mathbf{0}$, or strict inequality in the sufficient conditions (15-29) or (15-33). This is simply a statement that a steady state which is asymptotically stable to any disturbance must be unique.

BIBLIOGRAPHICAL NOTES

The first application of Liapunov's direct method to a catalytic reaction was by Wei,

> Wei, J., *Chem. Eng. Sci.*, *20*, 729 (1965).

Inequality (15-8) and some extensions are in

> Berger, A. J., and L. Lapidus, *AIChE J.*, *14*, 558 (1968).

The case $\mathscr{L} \neq 1$ is considered in

> Padmanabhan, L., R. Y. K. Yang, and L. Lapidus, *Chem. Eng. Sci.*, *26*, 1857 (1971).

This paper contains some errors. The variational approach in Sec. 15.3 follows

> Denn, M. M., *Chem. Eng. J. 4*, 105 (1972).

The subject is broadly reviewed in the books by Aris and Perlmutter,

> Aris, R., *The Mathematical Theory of Diffusion and Reaction in Permeable Catalysts*, Oxford University Press, Inc., New York, 1974.
> Perlmutter, D. D., *Stability of Chemical Reactions*, Prentice-Hall, Englewood Cliffs, N.J., 1972.

Both authors discuss approximations which reduce the analysis to one for lumped parameter systems. Perlmutter deals extensively with the related problem of the adiabatic tubular reactor. Additional references for chemical reactors are tabulated in

> Denn, M. M., in V. W. Weekman, Jr., ed., *Annual Review of Industrial and Engineering Chemistry, 1970,* American Chemical Society, Washington, D.C., 1972.

Section 15.4 follows a classic paper by Serrin on stability of solutions of the Navier-Stokes equations, including Poiseuille flow:

> Serrin, J., *Arch. Rat. Mech. Anal., 3,* 1 (1959).

Some extensions to viscoelastic liquids have been carried out in

> Feinberg, M. R., and W. R. Schowalter, *Proc. 5th Intern. Congr. Rheol., 1,* 201 (1969); *Ind. Eng. Chem. Fundamentals, 8,* 332 (1969).
> Kline, K. A., *Trans. Soc. Rheol., 14,* 335 (1970).
> Shahinpoor, M., and G. Ahmadi, *Arch. Rat. Mech. Anal., 47,* 188 (1972).

The contributions of Joseph and his coworkers are treated in his book

> Joseph, D. D., *Global Stability of Fluid Motion,* Springer-Verlag New York, Inc., New York, 1974.

The particular results cited on rotational Couette flow and thermal convective motion are in, respectively,

> Joseph, D. D., and W. Hung, *Arch. Rat. Mech. Anal., 44,* 1 (1971).
> Shir, C. C., and D. D. Joseph, *Arch. Rat. Mech. Anal., 30,* 38 (1968).

The latter paper deals in fact with simultaneous temperature and concentration gradients.

The application of Liapunov's direct method to distributed parameter systems has received considerable attention in recent years. General formulations are in

> Hahn, W., *Theory and Application of Liapunov's Direct Method,* Prentice-Hall, Englewood Cliffs, N.J., 1963.
> Zubov, V. I., *Methods of A. M. Lyapunov and Their Application,* Noordhoff, Groningen, Netherlands, 1964.

The connection with thermodynamics is discussed in

> Glansdorff, P., and I. Prigogine, *Thermodynamic Theory of Structure, Stability, and Fluctuations,* Wiley, New York, 1971.

Applications in control are tabulated in

> Wang, P. K. C., *Intern. J. Control, 7,* 101 (1968).

There are applications in several fields of continuum mechanics in papers in

Leipholz, H., ed., *Instability of Continuous Systems*, Springer, Berlin, 1971.

See also the review paper on thermoelastic stability,

Nemat-Nasser, S., *Appl. Mech. Rev.*, 23, 615 (1971).

Methods of constructing Liapunov functions for distributed systems are reviewed and developed in

Lin, Y. H., E. Kinnen, and J. C. Friedly, *Preprints 1973 Joint Automatic Control Conference*, I.E.E.E., New York, 1973.

Nonlinear Estimates by Averaging

16

16.1 Introduction

In several of the physical problems which we have examined the instability rapidly carries the system to a new equilibrium state. J. T. Stuart has developed an approximate method for estimating the new equilibrium state. The method gives accurate results when the new state is steady in time and spatially periodic, and the instability is characterized by the principle of exchange of stabilities (the eigenvalue is real at transition). The torque curve in rotational Couette flow and the Nusselt number in convective transport have been calculated by this method for flows beyond the point of instability. Stuart's method is a straightforward extension to distributed parameter systems of the averaging method in Sec. 6.2, except that we shall now be averaging over periodic spatial behavior rather than periodicity in time.

16.2 Averaged Equations

We shall demonstrate the averaging method for the special case of rotational Couette flow with the inner cylinder rotating and the outer fixed. As shown in Chapter 10, instability occurs when the Taylor number reaches a critical value of 3390, followed by a steady periodic secondary flow with wave number 3.12. The starting point is the steady-state, dimensionless Navier-Stokes and continuity equations, Eqs. (10-5) and (10-6), neglecting the body force term:

$$\mathbf{v}\cdot\nabla\mathbf{v} = -\nabla p + \frac{1}{\mathrm{Re}}\,\nabla^2\mathbf{v} \tag{16-1}$$

$$\nabla\cdot\mathbf{v} = 0 \tag{16-2}$$

The flow is periodic in the axial (z) direction and has no angular (θ) dependence.

Following the analysis in Sec. 6.2, it will be convenient to separate the flow into a part which is averaged over one wavelength in the axial direction, $\bar{\mathbf{v}}(r)$, and a periodic deviation from the average, which we denote simply as $\mathbf{u}(r, z)$. $\bar{\mathbf{v}}(r)$ corresponds to $\mathcal{A}_0$ in Eq. (6-11a). The pressure is similarly broken up, and we have

$$\mathbf{v}(r, z) = \bar{\mathbf{v}}(r) + \mathbf{u}(r, z) \tag{16-3a}$$

$$p(r, z) = \bar{p}(r) + y(r, z) \tag{16-3b}$$

Equations (16-1) and (16-2) are then written as

$$\bar{\mathbf{v}}\cdot\nabla\bar{\mathbf{v}} + \mathbf{u}\cdot\nabla\bar{\mathbf{v}} + \bar{\mathbf{v}}\cdot\nabla\mathbf{u} + \mathbf{u}\cdot\nabla\mathbf{u} = -\nabla\bar{p} - \nabla y + \frac{1}{\mathrm{Re}}\,\nabla^2\bar{\mathbf{v}} + \frac{1}{\mathrm{Re}}\,\nabla^2\mathbf{u} \tag{16-4}$$

$$\nabla\cdot\bar{\mathbf{v}} + \nabla\cdot\mathbf{u} = 0 \tag{16-5}$$

$\mathbf{u}$ and y average to zero over one wavelength in the axial direction, as do their derivatives. Averaging Eqs. (16-4) and (16-5) then leads to the equations

$$\bar{\mathbf{v}}\cdot\nabla\bar{\mathbf{v}} = -\nabla\bar{p} + \frac{1}{\mathrm{Re}}\,\nabla^2\bar{\mathbf{v}} - \overline{\mathbf{u}\cdot\nabla\mathbf{u}} \tag{16-6}$$

$$\nabla\cdot\bar{\mathbf{v}} = 0 \tag{16-7}$$

The term $\overline{\mathbf{u}\cdot\nabla\mathbf{u}}$ is the average of the product $\mathbf{u}\cdot\nabla\mathbf{u}$ taken over one wavelength. This term is analogous to the *Reynolds stresses* which occur in turbulent flow. In the absence of Reynolds stresses Eqs. (16-6) and (16-7) would be identical to the full Navier-Stokes and continuity equations, (16-1) and (16-2). Using the small gap assumption introduced in Sec. 10.2 the θ component of $\bar{\mathbf{v}}$, which we denote as $\overline{V}$, satisfies the equation

$$\frac{d^2\overline{V}}{dx^2} = \left(\frac{\delta}{R}\right)\mathrm{Re}\,\frac{d}{dx}\,\overline{uv} \tag{16-8}$$

x is the dimensionless radial distance measured outward from the surface of the rotating inner cylinder. $x = 1$ is the stationary outer cylinder. u, v, and w are, respectively, the radial, angular, and axial components of $\mathbf{u}$. The solution to Eq. (16-8) satisfying the no-slip boundary conditions is

$$\overline{V} = 1 - x + \left(\frac{\delta}{R}\right) \text{Re} \left[\int_0^x \overline{uv}\, dx - x \int_0^1 \overline{uv}\, dx\right] \qquad (16\text{-}9)$$

The first group of terms, $1 - x$, is the simple Couette solution with circular streamlines, Eq. (10-11), in the limit $\delta/R \to 0$. The second term represents the distortion of the mean flow which is caused by the secondary motion. Equation (16-9) is analogous to Eqs. (6-15) and (6-32).

Continuing the analogy to Secs. 6.2 and 6.4, we now form the inner product between $\mathbf{u}$ and Eq. (16-4). Using Eq. (16-6) this inner product can be manipulated into the form

$$0 = \mathbf{u}\cdot\overline{\mathbf{D}}\cdot\mathbf{u} - \frac{1}{\text{Re}}\mathbf{u}\cdot\nabla^2\mathbf{u} + \nabla\cdot[y\mathbf{u} + \frac{1}{2}(\mathbf{u}\cdot\mathbf{u})(\overline{\mathbf{v}} + \mathbf{u})] - \mathbf{u}\cdot(\overline{\mathbf{u}\cdot\nabla\mathbf{u}}) \quad (16\text{-}10)$$

$$\overline{\mathbf{D}} = \frac{1}{2}[\nabla\overline{\mathbf{v}} + (\nabla\overline{\mathbf{v}})^T] \qquad (16\text{-}11)$$

We now integrate Eq. (16-10) over a volume $\mathscr{V}$, which is one wavelength in the axial direction and which extends from the inner to the outer cylinder. The term $\mathbf{u}\cdot(\overline{\mathbf{u}\cdot\nabla\mathbf{u}})$ vanishes because the average of $\mathbf{u}$ over one wavelength is zero. The divergence term $\nabla\cdot[y\mathbf{u} + \frac{1}{2}(\mathbf{u}\cdot\mathbf{u})(\overline{\mathbf{v}} + \mathbf{u})]$ converts to a surface integral with Green's theorem and this vanishes in much the same way as similar surface integrals in Sec. 15.4. We obtain, then,

$$0 = \int_{\mathscr{V}} [\mathbf{u}\cdot\overline{\mathbf{D}}\cdot\mathbf{u} - \frac{1}{\text{Re}}\mathbf{u}\cdot\nabla^2\mathbf{u}]\, d\mathscr{V} \qquad (16\text{-}12)$$

Equation (16-12) can be considered an integral equation for the unknown function $\mathbf{u}$. It is nearly identical to the Reynolds-Orr equation (15-24) developed from Liapunov's method. The two equations relate to completely different situations, however. The Reynolds-Orr equation defines the behavior of a deviation from a steady state which is a solution of the Navier-Stokes equation. Stuart's equation (16-12) defines the deviation from the mean motion, and the mean velocity $\overline{\mathbf{v}}$ is *not* a solution of the Navier-Stokes equation.

The final working equation is obtained by writing Eq. (16-12) in terms of components. It can be shown by ordering arguments, or, more simply, by direct a posteriori calculation, that derivatives of v are negligible compared to derivatives of u and w. This approximation simplifies the $\mathbf{u} \cdot \nabla^2 \mathbf{u}$ term in Eq. (16-12) and we obtain, finally,

$$0 = \int_{x=0}^{1} \int_{z=0}^{h} \left\{ \frac{d\bar{V}}{dx} \overline{uv} + \left(\frac{R}{\delta}\right) \frac{1}{\text{Re}} \left[\frac{\partial u}{\partial z} - \frac{\partial w}{\partial x}\right]^2 \right\} dz \, dx \qquad (16\text{-}13)$$

z is made dimensionless here with the gap spacing, δ, and h is the dimensionless wavelength of the periodic motion in units of δ. We noted in Chapter 10 that h is experimentally very close to 2 in value.

16.3 Torque Equation

The functions u, v, and w are periodic in z, so they may be expanded in a Fourier series. For example,

$$v = -\mathscr{A} \sum_{n=1}^{\infty} \psi_n(x) \sin nkz$$

The minus sign and the amplitude, $\mathscr{A}$, are factored out for later convenience. k is the dimensionless wave number, $2\pi/h$. In analogy to the treatment in Chapter 6, we make the first of two basic assumptions here. We assume that only the term $n = 1$ contributes significantly to the flow, and that all higher harmonics may be neglected. With that assumption we write

$$u = -\frac{\mathscr{A}}{\text{Re}} \phi(x) \cos kz \qquad (16\text{-}14a)$$

$$v = -\mathscr{A}\psi(x) \sin kz \qquad (16\text{-}14b)$$

$$w = -\frac{\mathscr{A}}{k^2 \, \text{Re}} \frac{d\phi}{dx} \cos kz \qquad (16\text{-}14c)$$

The factor $1/\text{Re}$ is introduced into the expression for u to facilitate comparison with the linear theory in Chapter 10. The form of w is obtained from u by use of the continuity equation. Equations (16-9), (16-13), and (16-14) now combine to

$$0 = \int_0^1 \left[-1 + \frac{\mathscr{A}^2\delta}{R} S(x) \right] \phi(x)\psi(x) \, dx + \frac{2}{Tk^2} \int_0^1 [\psi'' - k^2\psi]^2 \, dx \qquad (16\text{-}15)$$

T is the Taylor number introduced in Sec. 10.4, and the function $S(x)$ is given by the equation

$$S(x) = \phi(x)\psi(x) - \int_0^1 \phi(x)\psi(x)\, dx \qquad (16\text{-}16)$$

The *critical Taylor number*, T_c, at which the instability occurs, must correspond to $\mathscr{A} = 0$, for until $T = T_c$ the streamlines are circular with $u = v = w = 0$. Equation (16-15) can then be solved for T_c at the point of transition,

$$T_c = \frac{2 \int_0^1 [\psi'' - k^2 \psi]^2\, dx}{k^2 \int_0^1 \phi(x)\psi(x)\, dx} \qquad (16\text{-}17)$$

and we finally obtain an expression for $\mathscr{A}$ by combining Eqs. (16-15) and (16-17) and rearranging,

$$\mathscr{A}^2 = \frac{R}{\delta}\left[1 - \frac{T_c}{T}\right] \frac{\int_0^1 \phi(x)\psi(x)\, dx}{\int_0^1 S(x)\phi(x)\psi(x)\, dx} \qquad (16\text{-}18)$$

The torque G required to turn the inner cylinder is given by the dimensional equation

$$G = \int_0^L \int_0^{2\pi} \tilde{\tau}_{r\theta} R^2\, d\theta\, d\tilde{z}\,\Big|_{x=0} = \frac{2\pi\mu R^3 \Omega L}{\delta} \frac{d\bar{V}}{dx}\Big|_{x=0}$$

L is the length of the cylinders. From Eqs. (16-9), (16-14), and (16-18) this can be written as

$$\frac{G\delta}{2\pi\mu R^3 \Omega L} = -1 - \frac{\left[\int_0^1 \phi(x)\psi(x)\, dx\right]^2}{\left[\int_0^1 S(x)\phi(x)\psi(x)\, dx\right]}\left[1 - \frac{T_c}{T}\right] \qquad (16\text{-}19)$$

This equation applies for $T \geq T_c$. For $T < T_c$ the dimensionless torque is given by the first term, -1, which contains the usual linear dependence of torque on viscosity. The extra torque from the additional motion varies linearly with $1 - T_c/T$.

The coefficient of $1 - T_c/T$ in Eq. (16-19) is evaluated by making the second major assumption in the analysis. At the point of instability the functions $\phi(x)$ and $\psi(x)$ are given by the linearized perturbation equations developed in Chapter 10. We assume that the *shapes* of these functions are independent of T for rotational speeds beyond the critical Taylor number. [This assumption is really implicit in the derivation of Eq. (16-18) from (16-15).] The magnitudes may be set arbitrarily, since ϕ and ψ enter Eq. (16-19) in such a way that the torque is independent of the magnitude and depends only on the shape. Thus, $\psi(x)$ and $\phi(x)$ are given by Eqs. (10-32) and (10-33), respectively, with the coefficients $C_2/C_1 = 0.0039$, $C_3/C_1 = -0.0011, \ldots$ computed from the linear analysis. C_1 reflects an arbitrary scale factor. When these functions are substituted into Eq. (16-18) the coefficient is found to have a value of 1.5 to two significant figures,

$$\frac{G\delta}{2\pi\mu R^3 \Omega L} = -1 - 1.5\left[1 - \frac{T_c}{T}\right] \qquad (16\text{-}20)$$

Figure 16.1 shows torque data for a solution of 60% glycerine-40% water with $R/\delta = 28$. There is some scatter in the data, reflected in part by a

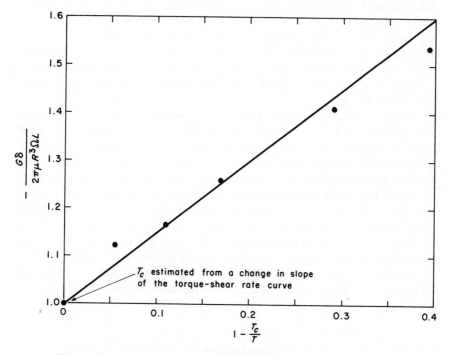

Figure 16.1.   Torque curve following the onset of instability between rotating cylinders for a 60% glycerine in water solution. These are the data from Fig. 1.5. [After M. M. Denn, Z.-S. Sun, and B. D. Rushton, *Trans. Soc. Rheol. 15*, 415 (1971), with permission.]

standard deviation of about 5% in estimating T_c in the series of experiments of which these data are a part. The trend of the data is clearly consistent with the theoretical prediction, which, it must be remembered, contains no adjustable parameters. The assumptions in the analysis break down as T exceeds T_c by about 20 or 30%, and at higher rotational speeds the theory overestimates the actual torque by a considerable amount. Other data reflect the same agreement between theory and experiment.

16.4 Non-Newtonian Fluid

The analysis can be extended in a straightforward manner to idealized non-Newtonian fluids, and the torque curve can be related to the normal stress functions $\tau_{rr} - \tau_{\theta\theta}$ and $\tau_{\theta\theta} - \tau_{zz}$. A reduction in torque is predicted because of the fluid elasticity. This is in quantitative agreement with experiment.

The estimate of the secondary flow calculated here, with its extension to simple viscoelastic fluids, has been used to partially explain the anomalous behavior of polymer solutions in rotational Couette flow. We noted in Sec. 10.5 that linear stability theory seems to predict the critical Taylor number accurately for polymer solutions but that the predicted critical cell size is considerably larger than the cell size actually observed. This is in contrast to Newtonian and inelastic non-Newtonian fluids, where the critical wave number of linear theory accurately predicts the cell spacing in the new equilibrium motion for $T > T_c$.

With an estimate of the magnitude of the secondary flow as given by Eq. (16-18) it is possible to calculate the energy of the motion. The energy is taken to be the sum of kinetic energy and entropic free energy associated with the alignment of polymer chains in solution under shear. For Newtonian and inelastic non-Newtonian fluids this energy is a minimum at the wave number corresponding to the critical value in linear theory, so it is reasonable to expect the cell spacing in the finite-amplitude motion to reflect the results of the linear calculation. The agreement seems to be coincidence, however. For a viscoelastic liquid the minimum energy will occur at a cell spacing which is lower than the critical value calculated from linear theory. In that case it is reasonable to expect the flow field following instability to adjust to a spacing which is closer than that computed from the most critical linear mode, in agreement with experiment.

BIBLIOGRAPHICAL NOTES

The approach developed here was first presented in

Stuart, J. T., *J. Fluid Mech.*, 4, 1 (1958).

Extensive comparisons with data are in

Donnelly, R. J., and N. J. Simon, *J. Fluid Mech.*, *7*, 401 (1960).

The comparison in Fig. 16.1 and the extension to non-Newtonian fluids are in

Denn, M. M., Z.-S. Sun, and B. D. Rushton, *Trans. Soc. Rheol.*, *15*, 415 (1971).

An application of the results of the latter study for evaluating dilute polymer solution rheological properties is

Jones, W. M., D. M. Davies, and M. C. Thomas, *J. Fluid Mech.*, *60*, 19 (1973).

The energy analysis for the cell spacing is in

Sun, Z.-S., and M. M. Denn, *AIChE J.*, *18*, 1010 (1972).

The analogous treatment for convective heat transport is in Appendix I of

Chandrasekhar, S., *Hydrodynamic and Hydromagnetic Stability*, Oxford University Press, Inc., New York, 1961.

The approximations in the analysis have been scrutinized by Davey, using the rigorous perturbation theory developed in the next chapter,

Davey, A., *J. Fluid Mech.*, *14*, 336 (1962).

Finite-Amplitude
Instability

17

17.1 Introduction

Our final task with respect to distributed parameter systems is to develop systematic analytical tools for estimating the response to finite-amplitude disturbances. In contrast to the Liapunov methods, which seek a priori bounds for stability without a detailed examination of the system dynamics, we seek here to describe those dynamics. The result is essentially the converse of the Liapunov approach, for we obtain estimates of the onset of instability. This will be demonstrated by focusing on two problems which we have studied previously. For a catalyst particle we shall obtain an estimate of the temperature perturbation which causes instability for a steady state which is stable to infinitesimal perturbations. For plane Poiseuille flow we shall estimate the dependence of the transition Reynolds number on the intensity of flow field fluctuations.

The analytical procedures which we shall develop in this chapter are straightforward generalizations of the Eckhaus method (Sec. 5.7) and the cascade method (Sec. 6.6). They are perturbation methods which depend on some kind of continiuity of behavior from linear to nonlinear regions. The phenomena of interest may occur beyond the range of validity of the approximations, and in some cases modes of behavior in the nonlinear regions

202

which do not have linear counterparts may go undetected. Nevertheless, these methods and variants of them offer the only analytical alternatives to direct numerical solution of the nonlinear partial differential equations.

17.2 Eckhaus Method

We shall develop the Eckhaus method for a catalytic reaction with unity Lewis number. For that case we showed in Sec. 9.3 that the adiabatic perturbation is the worst perturbation for linear theory. Since we are seeking a sufficient condition for instability, it suffices to assume the adiabatic perturbation here as well. Starting from Eq. (9-10) and expanding in the temperature deviation the dynamics are then described to *second order* by the partial differential equation

$$\frac{\partial \eta}{\partial t} = \frac{\partial^2 \eta}{\partial z^2} + \alpha F'(y_s)\eta + \frac{1}{2}\alpha F''(y_s)\eta^2 + \cdots \tag{17-1}$$

$$\frac{\partial \eta}{\partial z} = 0 \quad \text{at } z = 0, \quad \eta = 0 \quad \text{at } z = 1 \tag{17-2}$$

Stability to infinitesimal disturbances is determined by the linear eigenvalue problem (9-16) and (9-18):

$$\phi'' + [-\lambda + \alpha F'(y_s)]\phi = 0 \tag{17-3}$$

$$\phi'(0) = \phi(1) = 0 \tag{17-4}$$

We have established that the linear system is self-adjoint, so λ is real and an oscillatory response is not possible. It is further established in Appendix D.3 that the eigenfunctions are orthogonal, and we may take them to be normalized to unity. Thus, eigenfunctions $\phi_n(z)$ and $\phi_m(z)$ satisfy the relation

$$\int_0^1 \phi_n(z)\phi_m(z)\,dz = \begin{cases} 0, & n \neq m \\ 1, & n = m \end{cases} \tag{17-5}$$

In all that follows it is assumed that λ is negative and that the system is stable to infinitesimal disturbances.

Following the procedure outlined in Sec. 5.7 we expand the solution to Eq. (17-1) in the eigenfunctions of Eq. (17-3):

$$\eta(z, t) = \sum_{n=1}^{\infty} \mathcal{A}_n(t)\phi_n(z) \tag{17-6}$$

We now substitute into Eq. (17-1), multiply the entire equation by $\phi_p(z)$, integrate with respect to z from zero to 1, and apply Eq. (17-5) to obtain an infinite set of differential equations for the Fourier coefficients, $\mathscr{A}_p$:

$$\dot{\mathscr{A}}_p = \lambda_p \mathscr{A}_p + \sum_{m,\,n=1}^{\infty} I_{pmn} \mathscr{A}_m \mathscr{A}_n + \cdots \tag{17-7}$$

$$I_{pmn} = \frac{1}{2}\,\alpha \int_0^1 F''(y_s(z))\phi_p(z)\phi_m(z)\phi_n(z)\,dz \tag{17-8}$$

Equation (17-7) is identical to Eq. (5-70), and the analysis follows exactly the same steps as for the lumped system. The key assumption is that we are dealing with cases close to marginal stability, $\lambda_1 \to 0$. In that case the system of differential equations (17-7) is dominated by the single equation

$$\dot{\mathscr{A}}_1 = \lambda_1 \mathscr{A}_1 + I_{111}\mathscr{A}_1^2 + \cdots \tag{17-9}$$

with the solution

$$\mathscr{A}_1(t) = \frac{\mathscr{A}_1(0)e^{\lambda_1 t}}{1 + \mathscr{A}_1(0)[I_{111}/\lambda_1][1 - e^{\lambda_1 t}]} \tag{17-10}$$

For a stable steady state, $\lambda_1 < 0$, sufficiently small finite disturbances will decay. However, when a disturbance exceeds a critical magnitude

$$\mathscr{A}_1 > \mathscr{A}_{1c} = \left|\frac{\lambda_1}{I_{111}}\right| = \left|\frac{2\lambda_1}{\alpha \int_0^1 F''(y_s(z))\phi_1^3(z)\,dz}\right| \tag{17-11}$$

the disturbance will grow.

One of the critical assumptions in the Eckhaus method is that the eigenvalues are widely spaced. We noted for lumped systems that that assumption will rarely be applicable. For the distributed catalyst particle, however, and for many other distributed systems, the assumption is excellent, and the method is widely applicable. Some generalization along the lines outlined in Sec. 5.7 is necessary when the system is not self-adjoint, and further straightforward modification is needed when the eigenvalues are complex.

17.3 Power Series Method

The cascade method, as developed in Sec. 6.6, applies to systems with oscillatory time behavior. The method simplifies considerably when oscillations are not possible, and one of the basic manipulations in the application to partial differential equations is clearly revealed. Thus, the approach outlined here has only limited application, but it is a nice illustration of one key step.

We start again with Eq. (17-1), but now we assume that the solution can be expressed as a series in powers of a function of time, $\mathscr{A}(t)$, with spatially dependent coefficients:

$$\eta(z, t) = \sum_{n=1}^{\infty} \eta_n(z)\mathscr{A}^n(t) \tag{17-12}$$

The functions $\eta_n(z)$ must satisfy the boundary conditions (17-4), and it is convenient to normalize $\eta_1(z)$,

$$\int_0^1 \eta_1^2(z)\,dz = 1 \tag{17-13}$$

The function $\mathscr{A}(t)$ is further assumed to evolve according to an ordinary differential equation,

$$\mathscr{\dot A} = \sum_{n=1}^{\infty} \mu_n \mathscr{A}^n \tag{17-14}$$

Equations (17-12) and (17-14) are substituted into Eq. (17-1), and the resulting equations are separated according to the power of $\mathscr{A}$. From the linear term in $\mathscr{A}$ we obtain

$$\eta_1'' + [-\mu_1 + 2F'(y_s)]\eta_1 = 0 \tag{17-15}$$

This is identical to Eq. (17-3). Thus, η_1 corresponds to the eigenfunction of linear stability and μ_1 to the eigenvalue. We identify them with the first mode, ϕ_1 and λ_1.

From the quadratic term we obtain the equation

$$\eta_2'' + [-2\lambda_1 + \alpha F'(y_s)]\eta_2 = \mu_2 \phi_1 - \tfrac{1}{2}\alpha F''(y_s)\phi_1^2 \tag{17-16}$$

Equation (17-16) is nonhomogeneous. It always has a solution as long as $2\lambda_1$ is not an eigenvalue of the homogeneous left-hand side. Since λ_1 is an eigenvalue, $2\lambda_1$ cannot be one. There is an exceptional case, however, as we approach the point of marginal stability, $\lambda_1 \to 0$. At that point the self-adjoint homogeneous equation has an eigensolution, $\phi_1(z)$, and a solution to the nonhomogeneous equation for $\eta_2(z)$ will exist if and only if the right-hand side is orthogonal to the eigenfunction. Thus, at the point of marginal stability existence of a solution requires

$$0 = \mu_2 \int_0^1 \phi_1^2(z)\,dz - \frac{1}{2}\alpha \int_0^1 F''(y_s(z))\phi_1^3(z)\,dz$$

or, using Eqs. (17-8) and (17-13),

$$\mu_2 = I_{111} \tag{17-17}$$

Continuity requires that this value of μ_2 be used in the neighborhood of the point of marginal stability, so Eq. (17-14) becomes

$$\mathscr{A} = \lambda_1 \mathscr{A} + I_{111}\mathscr{A}^2 + \cdots \tag{17-18}$$

which is identical to Eq. (17-9). Thus, from an apparently quite different starting point we obtain the same result as with the Eckhaus method. The key conceptual step here is the orthogonality relation used in passing from Eq. (17-16) to Eq. (17-17).

17.4 Plane Poiseuille Flow

We can now consider the problem of the stability of plane Poiseuille flow to finite-amplitude disturbances in order to illustrate the application of the cascade method introduced in Sec. 6.6. Following the linear stability analysis in Chapter 11 we shall assume that there are velocity components only in the direction of mean flow (x direction) and the direction normal to the walls (y direction). In that case it is convenient to rewrite the dimensionless Navier-Stokes and continuity equations, (11-1) and (11-2), in an equivalent form known as the *two-dimensional vorticity equation*,

$$\left[\frac{\partial}{\partial t} + \frac{\partial \Psi}{\partial y}\frac{\partial}{\partial x} - \frac{\partial \Psi}{\partial x}\frac{\partial}{\partial y}\right]\left[\frac{\partial^2 \Psi}{\partial x^2} + \frac{\partial^2 \Psi}{\partial y^2}\right] - \frac{1}{\mathrm{Re}}\left[\frac{\partial^4 \Psi}{\partial x^4} + 2\frac{\partial^4 \Psi}{\partial x^2\,\partial y^2} + \frac{\partial^4 \Psi}{\partial y^4}\right] = 0$$

$$\tag{17-19}$$

The *stream function*, $\Psi(x, y)$, is related to the velocity components through the equations

$$v_x = \frac{\partial \Psi}{\partial y} \qquad v_y = -\frac{\partial \Psi}{\partial x} \tag{17-20}$$

The steady state, corresponding to the parabolic velocity profile in Eq. (11-4), is

$$\Psi_s = y - \tfrac{1}{3}y^3 \tag{17-21}$$

We now follow the procedure outlined in Sec. 6.6. In analogy to Eq. (6-53) the stream function is expanded in an exponential Fourier series,

$$\Psi(x, y, t) = \sum_m [\Psi_m e^{im\Theta} + \Psi_m^* e^{-im\Theta}] \tag{17-22}$$

where the Fourier coefficients Ψ_m are functions of y. Ψ^* is the complex conjugate of Ψ. Θ is a time coordinate moving with the fluid,

$$\Theta = k(x - c_R t) \tag{17-23}$$

so Eq. (17-22) represents a decomposition into a series of traveling waves. We have defined the wave velocity as c_R because in the linear limit it must correspond to the wave velocity of linear theory introduced in Sec. 11.2. The wave number, k, is introduced to retain the analogy in form to Eq. (11-9). For $n \geq 1$ the no-slip boundary conditions require that the Fourier coefficients satisfy

$$\Psi_m = \frac{\partial \Psi_m}{\partial y} = 0 \qquad \text{at } y = \pm 1 \tag{17-24}$$

The Fourier coefficients and the wave number depend on the time-dependent amplitude of the disturbance velocity, $\mathscr{A}(t)$. Following Eq. (6-54) the Fourier coefficients are expanded as power series in $\mathscr{A}$,

$$\Psi_m = \sum_{n \geq m} \Psi_{mn}(y) \mathscr{A}^n \tag{17-25}$$

The functions $\Psi_{mn}(y)$ must all satisfy the boundary conditions (17-24). The logic for taking the sum over $n \geq m$ has already been discussed in Sec. 6.6. In Eq. (17-25) we have omitted for convenience the powers of 2 introduced into Eq. (6-56). $\mathscr{A}$ and c_R evolve according to differential equations:

$$\dot{\mathscr{A}} = k[a_0 \mathscr{A} + a_1 \mathscr{A}^2 + a_2 \mathscr{A}^3 + \cdots] \tag{17-26}$$

$$-\frac{d}{dt} c_R t = b_0 + b_1 \mathscr{A} + b_2 \mathscr{A}^2 + \cdots \tag{17-27}$$

Equations (17-26) and (17-27) correspond to Eqs. (6-55) and (6-56), except for the omitted powers of 2. The factor of k is introduced into Eq. (17-26) so that a_0 corresponds to c_I from linear theory, while the minus sign in Eq. (17-27) is for convenience in comparison with the published literature. $-b_0$ is the linear theory value for c_R.

Equations (17-22) through (17-27) are now substituted into Eq. (17-19) and the terms are grouped according to Fourier component (m) and power of $\mathcal{A}(n)$. The coefficient of each m, n term is then set to zero, resulting after tedious calculation in a set of equations for the functions $\Psi_{mn}(y)$. It follows at once that $2\Psi_{00} = \Psi_s$, Eq. (17-21). Using the notation

$$c_n = ia_n - b_n$$

we obtain, in addition, the following equations:

$$\Psi_{11}^{IV} - 2k^2\Psi_{11}'' + k^4\Psi_{11} - ik \operatorname{Re}[(1 - y^2 - c_0)(\Psi_{11}'' - k^2\Psi_{11}) + 2\Psi_{11}] = 0$$
$$(17\text{-}28)$$

$$\Psi_{02}^{IV} - 2ka_0 \operatorname{Re}\Psi_{02}'' = k \operatorname{Re}\operatorname{Imag}\{\Psi_{11}'\Psi_{11}^{*\prime\prime} + \Psi_{11}\Psi_{11}^{*\prime\prime\prime}\} \qquad (17\text{-}29)$$

$$\Psi_{22}^{IV} - 2k^2\Psi_{22}'' + 4k^4\Psi_{22} - 2ik \operatorname{Re}[(+1 - y^2 - c_0)(\Psi_{22}'' - 4k^2\Psi_{22}) + 2\Psi_{22}]$$
$$= ik \operatorname{Re}(\Psi_{11}'\Psi_{11}'' - \Psi_{11}\Psi_{11}''') \quad (17\text{-}30)$$

$$\Psi_{13}^{IV} - 2k^2\Psi_{13}'' + k^4\Psi_{13} - ik \operatorname{Re}[(1 - y^2 - c_0 - 2ia_0)(\Psi_{13}'' - k^2\Psi_{13}) + 2\Psi_{13}]$$
$$= -ikc_2 \operatorname{Re}(\Psi_{11}'' - k^2\Psi_{11}) - i \operatorname{Re}[-2\Psi_{02}'(\Psi_{11}'' - k^2\Psi_{11})$$
$$- 2\Psi_{11}^{*\prime}(\Psi_{22}'' - 4k^2\Psi_{22}) + \Psi_{22}'(\Psi_{11}^{*\prime\prime} - k^2\Psi_{11}^{*}) + 2\Psi_{11}\Psi_{02}'''$$
$$- \Psi_{11}^{*}(\Psi_{22}''' - 4k^2\Psi_{22}') + 2\Psi_{22}(\Psi_{11}^{*\prime\prime\prime} - k^2\Psi_{11}^{*\prime})]$$
$$\equiv -ikc_2 \operatorname{Re}(\Psi_{11}'' - k^2\Psi_{11}) - i \operatorname{Re}\mathcal{H}_{13}(\Psi_{02}, \Psi_{11}, \Psi_{22}) \qquad (17\text{-}31)$$

Imag{ } in Eq. (17-29) denotes the imaginary part of the complex function. We have already made use of the fact, which can be shown from the complete expansion, that $\Psi_{mn} = 0$ when $m + n$ equals an odd integer, and all odd values of c_n are zero.

This set of equations can be solved sequentially. Equation (17-28) is simply the Orr-Sommerfeld equation of linear theory, (11-15), and c_0 is the eigenvalue whose calculation is discussed in Sec. 11.5. Equations (17-29) and (17-30) are nonhomogeneous linear equations whose solutions can be obtained when Ψ_{11} is available. Equation (17-30) has a structure which is similar to that of the Orr-Sommerfeld equation and requires the same kinds of specialized numerical analysis. In the case of Eq. (17-31) we have a situation exactly like that discussed in the preceding section. The homogeneous part (left-hand side) of the equation is nearly identical to the homogeneous Orr-Sommerfeld equation, differing only in the term $-2ia_0$. As long as $a_0 \neq 0$ the homogeneous part of the equation has no eigensolution, and a solution to the nonhomogeneous equation exists regardless of the value of c_2. At the curve of marginal stability, however, $a_0 = 0$, and Ψ_{11} is a solution to the homogeneous part of the equation. The full nonhomogeneous equation has a solution only under the special case that the right-hand side is orthogonal to the adjoint of the eigenfunction.

The adjoint for the Orr-Sommerfeld equation is developed in Appendix E as the solution to the equation

$$\Psi_{11}^{A\text{IV}} - 2k^2\Psi_{11}^{A''} + k^4\Psi_{11}^{A} - ik\,\text{Re}\,[(1-y^2-c_0)(\Psi_{11}^{A''} - k^2\Psi_{11}^{A}) - 4y\Psi_{11}^{A'}] = 0 \tag{17-32}$$

with the same boundary conditions as those on Ψ_{11}. The condition that Ψ_{11}^{A} be orthogonal to the right-hand side of Eq. (17-31) leads to an equation for c_2,

$$c_2 = ia_2 - b_2 = \frac{-\displaystyle\int_{-1}^{1}\Psi_{11}^{A}\,\mathscr{H}_{13}(\Psi_{02},\,\Psi_{11},\,\Psi_{22})\,dy}{k\displaystyle\int_{-1}^{1}\Psi_{11}^{A}(\Psi_{11}'' - k^2\Psi_{11})\,dy} \tag{17-33}$$

We shall take this as the defining equation for c_2 everywhere in order to ensure continuity at the neutral curve.

The coefficients in Eq. (17-26) are now all determined. a_0 is obtained as the growth rate in the linear theory, a_1 is zero, and a_2 is given by Eq. (17-33). The solution is given by

$$\mathscr{A}(t) = \frac{\mathscr{A}(0)e^{a_0 kt}}{[1 + (\mathscr{A}^2(0)/\mathscr{A}_\infty^2)(e^{2a_0 kt} - 1)]^{1/2}} \tag{17-34}$$

$$\mathscr{A}_\infty^2 = -\frac{a_2}{a_0} \tag{17-35}$$

When the system is stable to infinitesimal disturbances $a_0 < 0$. In that case $\mathscr{A}(t)$ will always decay to zero if $a_2 < 0$. If $a_2 > 0$, however, $\mathscr{A}(t)$ will grow without bound if $\mathscr{A}(0) > \mathscr{A}_\infty$. Thus, $\mathscr{A}_\infty$ defines the critical amplitude for instability to a finite disturbance. Just as in the strictly linear theory, we can compute the value of $\mathscr{A}_\infty$ for each pair k, Re. A line of constant $\mathscr{A}_\infty$ then defines a curve of neutral stability for a given amplitude, and the minimum of that curve defines the critical Reynolds number for a given amplitude. Such a plot is shown in Fig. 17.1, and the critical Reynolds number is shown as a function of $\mathscr{A}_\infty$ on the curve denoted $E = 0$ in Fig. 17.2. The value $\mathscr{A}_\infty = 0.004$ was found to be the limit of computational accuracy.

The curve in Fig. 17.2 shows the kind of behavior which we expect. As we increase the size of the velocity fluctuation the transition Reynolds number decreases. For purposes of interpretation it is helpful to note that centerline turbulence intensity, the ratio of the root mean square velocity fluctuation at the centerline to the mean centerline velocity, is equal to $\sqrt{2}\,k\mathscr{A}_\infty$. Thus, Fig. 17.2 corresponds to intensities of up to about 6×10^{-3}. Fluctuation

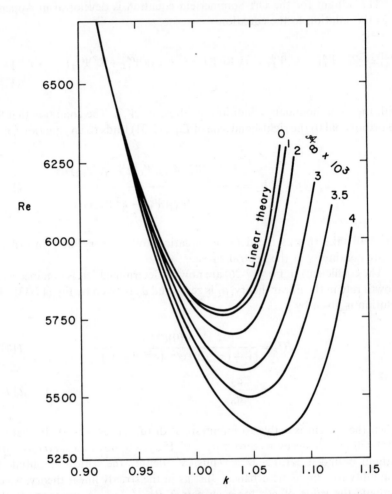

Figure 17.1. Curves of marginal stability for flow between parallel plates for various disturbance amplitudes. [After K. C. Porteous and M. M. Denn, *Trans. Soc. Rheol. 16*, 309 (1972), with permission].

intensities are of the order of several percent at the experimental transition to turbulence, so it is not surprising that the calculated values of the critical Reynolds number remain considerably above the experimental value of about 1000. Indeed, at higher amplitudes it is likely that other effects not considered in this two-dimensional perturbation analysis become important and may control the ultimate transition.

The same calculation can be carried out in a straightforward manner for

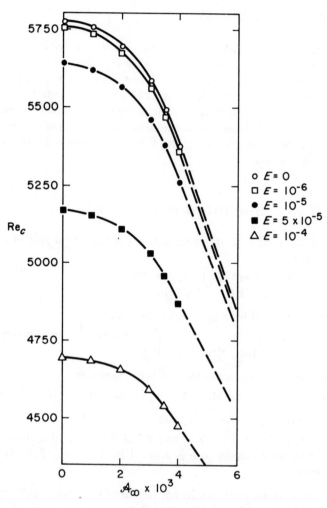

Figure 17.2. Critical Reynolds number as a function of disturbance amplitude for various values of elasticity number. [After K. C. Porteous and M. M. Denn, *Trans. Soc. Rheol.*, *16*, 309 (1972), with permission.]

dilute polymer solutions. The decrease in critical Reynolds number with disturbance amplitude is shown in Fig. 17.2 for values of E up to 10^{-4}. The elasticity number, E, is defined in Sec. 11.6. A careful examination of the data in Fig. 17.2 shows that in this dilute solution range the transition Reynolds numbers of Newtonian and slightly viscoelastic liquids come closer together with increasing disturbance amplitude.

17.5 Concluding Remarks

The nonlinear methods outlined in this chapter are quite tedious, but they are straightforward in principle and can provide a considerable amount of useful information without the necessity of direct simulation of the nonlinear partial differential equations. Both the Eckhaus and cascade formalisms, as well as other similar methods, have been applied to a number of problems in addition to the two discussed here. Analytical estimation of the behavior of nonlinear distributed systems is still far from well developed and remains an active area of research.

BIBLIOGRAPHICAL NOTES

The approaches used here are grounded in a fundamental pair of papers by Stuart and Watson,

>   Stuart, J. T., *J. Fluid Mech.*, *9*, 353 (1960).
>   Watson, J., *J. Fluid Mech.*, *9*, 371 (1960).

The subject is covered broadly in Eckhaus's book,

>   Eckhaus, W., *Studies in Non-Linear Stability Theory*, Springer-Verlag New York, Inc., New York, 1965.

The formalism for the cascade method is developed in

>   Reynolds, W. C., and M. C. Potter, *J. Fluid Mech.*, *27*, 465 (1967).

For a recent comprehensive survey, see

>   Stuart, J. T., in M. Van Dyke and W. G. Vincenti, eds., *Annual Review of Fluid Mechanics*, vol. 3, Annual Reviews, Inc., Palo Alto, Calif., 1971.

Two surveys focusing more on the convective heat transfer problem are

>   Segal, L. A., in R. J. Donnelly, R. Herman, and I. Prigogine, eds., *Non-Equilibrium Thermodynamics, Variational Techniques and Stability*, University of Chicago Press, Chicago, 1966.
>   Whitehead, J. A., *Am. Scientist*, *59*, 444 (1971).

Some recent foundation papers are

>   DiPrima, R. C., W. Eckhaus, and L. A. Segal, *J. Fluid Mech.*, *49*, 705 (1971).
>   Hocking, L. M., and K. Stewartson, *Proc. Roy. Soc. (London)*, *A326*, 289 (1972).
>   Iooss, G., *Arch. Rat. Mech. Anal.*, *40*, 166 (1971).

Joseph, D. D., and D. H. Sattinger, *Arch. Rat. Mech. Anal.*, *45*, 79 (1972).

Stewartson, K., and J. T. Stuart, *J. Fluid Mech.*, *48*, 529 (1971).

The treatment for the catalyst particle in Sec. 17.2 is from

Denn, M. M., *Chem. Eng. J.*, *4*, 105 (1972).

The development in Sec. 17.4 for plane Poiseuille flow was first done by Reynolds and Potter in the paper cited above, where they also studied combined Couette-Poiseuille flow. The calculations for Newtonian and viscoelastic liquids in Figs. 17.1 and 17.2 are by Porteous and were reported in

Porteous, K. C., and M. M. Denn, *Trans. Soc. Rheol.*, *16*, 309 (1972).

Similar calculations are in

Cousins, R. R., *Intern. J. Eng. Sci.*, *10*, 511 (1972)

McIntire, L. V., and C. H. Lin, *J. Fluid Mech.*, *52*, 273 (1972).

The Reynolds-Potter and Porteous-Denn results are compared with direct numerical solution of the transient Navier-Stokes equations in

George, W. D., J. D. Hellums, and B. Martin, *J. Fluid Mech.*, *63*, 765 (1974).

Postface

The goal of the book, as described in the Preface, has been to provide a sufficient introduction to linear and nonlinear stability to enable the reader to begin to attack his own problems and to read the current literature. I believe that that point has been reached. It is useful here, however, to record some additional literature references. These are to stability problems of scientific and engineering interest for which there was no convenient place in the main body of the text for a citation.

There are a number of stability problems which are important in various aspects of heat exchanger performance. These are analyzed and reviewed in several papers in the symposium proceedings

Heat Transfer-Tulsa, AIChE Symp. Ser., *68*, No. 118 (1972).

We have not considered flow problems in which the energy equation has a major influence on the basic flow because of viscosity variations with temperature. Nonunique steady states can exist in such flows, and dynamical instabilities are predicted. These analyses may be relevant to some flow instabilities in polymer processing in which a hot fluid is brought into contact with a cold die. See

Pearson, J. R. A., Y. T. Shah, and E. S. A. Vieira, *Chem. Eng. Sci.*, *28*, 2079 (1973); *29*, 1485 (1974).
Sukanek, P. C., C. A. Goldstein, and R. L. Lawrence, *J. Fluid Mech.*, *57*, 651 (1973).

The stability of the interface separating solid and liquid during crystal growth has been studied extensively using both classical thermodynamic arguments and dynamic stability analyses like those covered here. The thermodynamic arguments do not account for dynamics and can only establish instability. The field has been surveyed by Sekerka, who has been a major contributor:

Sekerka, R. F., *J. Crystal Growth, 3–4*, 71 (1968).

The most comprehensive study, including a nonlinear analysis, is by

Wollkind, D. J., and L. A. Segal, *Phil. Trans. Roy. Soc. (London), A268*, 351 (1970).

A similar moving boundary problem involving fog formation is in

Miller, C. A., and K. Jain, *Chem. Eng. Sci., 28*, 157 (1973).

The stability of the state of homogeneous fluidization in a fluidized bed was independently analyzed in

Ruckenstein, E., *Rev. Phys. Acad. Rep. Populaire Roumaine, 7*, 137 (1962).
Jackson, R., *Trans. Inst. Chem. Engrs. (London), 41*, 13 (1963).
Pigford, R. L., and T. Baron, *Ind. Eng. Chem. Fundamentals, 4*, 81 (1965).

More comprehensive models are considered in

Anderson, T. B., and R. Jackson, *Ind. Eng. Chem. Fundamentals, 7*, 12 (1968).
Drew, D. A., and L. Segal, *Studies Appl. Math., 3*, 233 (1971).

One mode by which a binary mixture breaks up into two phases is known as spinodal decomposition. This mechanism appears to be a consequence of small fluctuations over the entire volume, in contrast to the nucleation mechanism initiated by large localized fluctuations. The thermodynamic study of spinodal decomposition dates to J. Willard Gibbs. The modern dynamic theory is in

Cahn, J. W., *Acta. Met., 9*, 795 (1961); *Trans. Met. Soc. AIME, 242*, 166 (1968).

The latter paper is a survey. A similar analysis has been applied by Ruckenstein to explain heterogeneities in concentrated colloidal suspensions,

Ruckenstein, E., *J. Colloid Int. Sci.*, in press.

The development of pattern, or structure, from a homogeneous state is of considerable interest in theoretical biology. The idea that such structure can evolve because of stability considerations in a multicomponent system with diffusion and reaction was first demonstrated in a classic paper by Turing,

> Turing, A. M., *Phil. Trans. Roy. Soc. (London)*, *B237*, 37 (1952).

Extensions of this work are in

> Othmer, H. G., and L. E. Scriven, *Ind. Eng. Chem. Fundamentals*, *8*, 302 (1969); *J. Theoret. Biol.*, *32*, 507 (1971); *J. Theoret. Biol.*, in press.
>
> Glandsdorff, P., and I. Prigogine, *Thermodynamic Theory of Structure, Stability, and Fluctuation*, Wiley, New York, 1971.
>
> Segal, L. A., and J. L. Jackson, *J. Theoret. Biol.* 37, 595 (1972).

The term *dissipative structure* is sometimes used, since the patterns arise in a dissipative system.

Finally, the mathematics throughout the book have been maintained at the level usually covered in first graduate courses. Some researchers recently have begun to use the power of modern functional analysis both to solve specific stability problems and to unify general concepts. The mathematically prepared reader will find the following references useful as an introduction to this literature:

The basic results of functional analysis applicable to problems of the type considered in this book are developed in several texts by Krasnosel'skii,

> Krasnosel'skii, M. A., *Topological Methods in the Theory of Nonlinear Integral Equations*, Pergamon, New York, 1964.
>
> Krasnosel'skii, M. A., *Positive Solutions of Operator Equations*, Noordhoff, Groningen, Netherlands, 1964.
>
> Krasnosel'skii, M. A., *Translations Along Trajectories of Differential Equations*, American Mathematical Society, Providence, R.I., 1968.

The first of these is the most relevant. The main ideas are also contained in

> Berger, M. S., and M. S. Berger, *Perspectives in Nonlinearity: An Introduction to Nonlinear Analysis*, W. A. Benjamin, New York, 1968.

A number of review papers and specific applications are in

> Keller, J. B., and S. Antman, *Bifurcation Theory and Nonlinear Eigenvalue Problems*, W. A. Benjamin, New York, 1969.

A sequence of applications in fluid mechanics by Iudovich is instructive. Complete references may be found in

> Iudovich, V. I., *J. Appl. Math. and Mech.* (translation of *Prik. Mat. i Mekh.*), *31*, 103, 294 (1967).

See also the review paper

> Görtler, H., and W. Velte, *Physics of Fluids (Supplement)*, *10*, S3 (1967).

Applications to the equations of reaction engineering are in

Gavalas, G. R., *Nonlinear Differential Equations of Chemically Reacting Systems*, Springer-Verlag New York Inc., New York, 1968.

Cohen, D. S., and T. W. Laetsch, *J. Diff. Equations*, 7, 217 (1970).

Cohen, D. S., *SIAM J. Appl. Math.*, 20, 7 (1971).

Keller, H. B., *J. Math. and Mech.*, 19, 279, (1969); *J. Diff. Equations*, 7, 415 (1970).

Appendix A:
A Theorem
on Eigenvalues

We shall prove the following theorem, which was used in Sec. 3.3: *Let* **B** *be a symmetric, positive definite matrix. Then* **B·A** *is negative definite if and only if all eigenvalues of* **A** *have negative real parts.*

Sign definiteness is associated with a quadratic form, so only the symmetric part of the matrix, $\frac{1}{2}(\mathbf{B \cdot A} + \mathbf{A}^T \cdot \mathbf{B})$, need be considered. Thus, it is equivalent to prove the following: *Define symmetric matrices* **B** *and* **C** *by the equation*

$$\mathbf{B \cdot A} + \mathbf{A}^T \cdot \mathbf{B} = -\mathbf{C} \tag{A-1}$$

Positive definite matrices **B** *and* **C** *exist if and only if all eigenvalues of* **A** *have negative real parts.*

The "if" part of the theorem is proved by noting that a formal solution to Eq. (A-1) can be written as

$$\mathbf{B} = \int_0^\infty \exp(\mathbf{A}^T t) \cdot \mathbf{C} \cdot \exp(\mathbf{A}t) \, dt \tag{A-2}$$

218

$\exp(\mathbf{A}t)$ is the matrix exponential, which satisfies the matrix equation

$$\frac{d}{dt}\exp(\mathbf{A}t) = \mathbf{A}\cdot\exp(\mathbf{A}t)$$

The integral converges if the eigenvalues of $\mathbf{A}$ have negative real parts. That Eq. (A-2) is a solution of Eq. (A-1) is established by direct substitution:

$$\mathbf{B}\cdot\mathbf{A} + \mathbf{A}^T\cdot\mathbf{B} = \int_0^\infty \{\exp(\mathbf{A}^T t)\cdot\mathbf{C}\cdot\exp(\mathbf{A}t)\cdot\mathbf{A} + \mathbf{A}^T\cdot\exp(\mathbf{A}^T t)\cdot\mathbf{C}\cdot\exp(\mathbf{A}t)\}\,dt$$

$$= \int_0^\infty \frac{d}{dt}\{\exp(\mathbf{A}^T t)\cdot\mathbf{C}\cdot\exp(\mathbf{A}t)\}\,dt = -\mathbf{C}$$

We construct a quadratic form from Eq. (A-2) by pre- and postmultiplying a constant vector, $\mathbf{y}_0$:

$$\mathbf{y}_0\cdot\mathbf{B}\cdot\mathbf{y}_0 = \int_0^\infty \mathbf{y}(t)\cdot\mathbf{C}\cdot\mathbf{y}(t)\,dt \tag{A-3}$$

where $\mathbf{y}(t)$ is defined as

$$\mathbf{y}(t) = \exp(\mathbf{A}t)\cdot\mathbf{y}_0 \tag{A-4}$$

or

$$\dot{\mathbf{y}} = \mathbf{A}\cdot\mathbf{y}, \qquad \mathbf{y}(0) = \mathbf{y}_0 \tag{A-5}$$

For every positive definite matrix $\mathbf{C}$ it follows from Eq. (A-3) that there is a corresponding positive definite matrix $\mathbf{B}$, defined by Eq. (A-2). This proves the first half of the theorem.

"Only if" is proved by considering the quadratic form $\mathbf{y}\cdot\mathbf{C}\cdot\mathbf{y}$, where $\mathbf{y}$ is defined by Eq. (A-5). For positive definite $\mathbf{B}$ and $\mathbf{C}$ we have

$$\mathbf{y}\cdot\mathbf{C}\cdot\mathbf{y} = -(\mathbf{y}\cdot\mathbf{B}\cdot\mathbf{A}\cdot\mathbf{y} + \mathbf{y}\cdot\mathbf{A}^T\cdot\mathbf{B}\cdot\mathbf{y}) = -\frac{d}{dt}(\mathbf{y}\cdot\mathbf{B}\cdot\mathbf{y}) > 0$$

which can be written in an equivalent form

$$\mathbf{y}\cdot\mathbf{B}\cdot\mathbf{y} = \mathbf{y}_0\cdot\mathbf{B}\cdot\mathbf{y}_0 - \int_0^t \mathbf{y}\cdot\mathbf{C}\cdot\mathbf{y}\,dt > 0 \tag{A-6}$$

We now show that for an arbitrary $\mathbf{y}_0$, $\mathbf{y} = \exp(\mathbf{A}t)\cdot\mathbf{y}_0$ must come arbitrarily close to zero for sufficiently large t. The proof is by contradiction. If $\mathbf{y}$ does not come arbitrarily close to zero, then there must be some positive number N such that the integral in Eq. (A-6) is bounded by Nt and we can write

$$\mathbf{y} \cdot \mathbf{B} \cdot \mathbf{y} < \mathbf{y}_0 \cdot \mathbf{B} \cdot \mathbf{y}_0 - Nt \qquad (A\text{-}7)$$

For a sufficiently large value of t we must then have $\mathbf{y} \cdot \mathbf{B} \cdot \mathbf{y} < 0$, which contradicts the assumption that $\mathbf{B}$ is positive definite. Thus, by contradiction, $\exp(\mathbf{A}t) \cdot \mathbf{y}_0$ must come arbitrarily close to zero for sufficiently large t and arbitrary $\mathbf{y}_0$. This requires that the eigenvalues of $\mathbf{A}$ all have negative real part. Hence, the proof of the theorem is complete.

BIBLIOGRAPHICAL NOTES

The proof can be done in various ways. See

Bellman, R., *Introduction to Matrix Analysis*, McGraw-Hill, New York, 1960.

Gantmacher, F. R., *The Theory of Matrices*, Chelsea Publishing Company, Inc., New York, 1959.

Gantmacher, F. R., *Applications of the Theory of Matrices*, Wiley-Interscience, New York, 1959.

LaSalle, J., and S. Lefschetz, *Stability by Liapunov's Direct Method*, Academic Press, New York, 1961.

The book by Gantmacher is a two-volume set, translated from the Russian. *Applications* is a translation of the second volume.

Appendix B:
Perturbation
Solutions

We seek perturbation solutions to a nonlinear algebraic eigenvalue problem and its differential equation analog.

B.1 Algebraic Equation

Equation (5-35) can be written through second order in $|\mathbf{y}|$ as

$$\sum_j D_{ij} y_j = \lambda \left\{ \sum_j [\beta_{ij} + \beta_{ji}] y_j + \sum_{j,k} [\gamma_{ijk} + \gamma_{jki} + \gamma_{kij}] y_k y_j + \cdots \right\} \quad \text{(B-1)}$$

$$D_{ij} = C_{ij} + C_{ji}$$

$$\beta_{ij} = B_{ij}(0)$$

$$\gamma_{ijk} = \frac{\partial B_{jk}}{\partial \xi_i} \qquad \text{at } \xi = 0$$

221

We seek a solution in the form

$$y = \mathscr{A}\mathbf{y}_0 + \mathscr{A}^2\mathbf{z} + \cdots \tag{B-2}$$

$$\lambda = \lambda_0 + \mathscr{A}\lambda_1 + \cdots \tag{B-3}$$

$$|\mathbf{y}_0| = 1 \tag{B-4}$$

Substituting into Eq. (B-1) and equating coefficients of each power of $\mathscr{A}$ to zero we obtain the following sequence of linear equations:

$$\mathscr{A}^0: \quad \sum_j D_{ij} y_{0j} = \lambda_0 \sum_j [\beta_{ij} + \beta_{ji}] y_{0j} \tag{B-5}$$

$$\mathscr{A}^1: \quad \sum_j D_{ij} z_j = \lambda_0 \sum_j [\beta_{ij} + \beta_{ji}] z_j + \left\{ \lambda_1 \sum_j [\beta_{ij} + \beta_{ji}] y_{0j} \right.$$

$$\left. + \lambda_0 \sum_{j,k} [\gamma_{ijk} + \gamma_{jki} + \gamma_{kij}] y_{0k} y_{0j} \right\} \tag{B-6}$$

Equation (B-5) is a linear eigenvalue equation which, with the normalization (B-4), uniquely defines $\mathbf{y}_0$ and λ_0. This is identical to the system (5-41) and (5-43). The homogeneous part of Eq. (B-6) is identical to Eq. (B-5), indicating that $\mathbf{y}_0$ is an eigenvector. Since the homogeneous part has an eigensolution, there will be a solution to the nonhomogeneous equation if and only if the non-homogeneous part is orthogonal to the eigenvector; that is,

$$\lambda_1 \sum_{i,j} [\beta_{ij} + \beta_{ji}] y_{0i} y_{0j} + \lambda_0 \sum_{i,j,k} [\gamma_{ijk} + \gamma_{jki} + \gamma_{kij}] y_{0i} y_{0j} y_{0k} = 0 \tag{B-7}$$

The second sum is symmetric in i, j, and k, so we may write

$$\lambda_1 = -\lambda_0 \frac{3 \sum\limits_{i,j,k} \gamma_{ijk} y_{0i} y_{0j} y_{0k}}{\sum\limits_{i,j} [\beta_{ij} + \beta_{ji}] y_{0i} y_{0j}}$$

$$\lambda = \lambda_0 \left\{ 1 - \frac{3 \sum\limits_{i,j,k} \gamma_{ijk} y_{0i} y_{0j} y_{0k}}{\sum\limits_{i,j} [\beta_{ij} + \beta_{ji}] y_{0i} y_{0j}} \mathscr{A} + \cdots \right\} \tag{B-8}$$

B.2 Differential Equation

We seek a solution to Eq. (15-10) in the form

$$v = \mathscr{A}v_0 + \mathscr{A}^2 w + \cdots \tag{B-9}$$

$$\lambda = \lambda_0 + \mathscr{A}\lambda_1 + \cdots \tag{B-10}$$

$$\int_0^1 v_0^2(z)\, dz = 1 \tag{B-11}$$

Substituting into Eq. (15-10) and setting coefficients of each power of $\mathscr{A}$ to zero then leads to the following sequence of equations, where use has been made of the definition of $\phi(y_s, v)$ in Eq. (15-2):

$$\mathscr{A}^0: \quad v_0'' + \lambda_0 F'(y_s)v_0 = 0 \tag{B-12}$$

$$\mathscr{A}^1: \quad w'' + \lambda_0 F'(y_s)w = -\lambda_1 F'(y_s)v_0 - \tfrac{3}{4}\lambda_0 F''(y_s)v_0^2 \tag{B-13}$$

$$v_0'(0) = w'(0) = \cdots = v_0(1) = w(1) = \cdots = 0 \tag{B-14}$$

Equation (B-12) is identical to Eq. (15-12). Equation (B-13) has a homogeneous part identical to Eq. (B-12). Thus, λ_0 is an eigenvalue, v_0 an eigenfunction, and a nontrivial solution exists if and only if the right-hand side is orthogonal to v_0; that is,

$$0 = -\lambda_1 \int_0^1 F'(y_s)v_0^2\, dz - \frac{3}{4}\lambda_0 \int_0^1 F''(y_s)v_0^3\, dz \tag{B-15}$$

which, upon rearrangement and combination with Eq. (B-10), leads to the first-order approximation to the eigenvalue,

$$\lambda = \lambda_0 \left[1 - \frac{3}{4} \frac{\displaystyle\int_0^1 F''(y_s)v_0^3\, dz}{\displaystyle\int_0^1 F'(y_s)v_0^2\, dz} \mathscr{A} + \cdots \right] \tag{B-16}$$

BIBLIOGRAPHICAL NOTES

The basic theory of linear algebraic equations which is used in passing from Eq. (B-6) to (B-7) is covered in all texts on matrix theory. See, for example,

Amundson, N. R., *Mathematical Methods in Chemical Engineering: Matrices and Their Application*, Prentice-Hall, Englewood Cliffs, N.J., 1966.

The equivalent result in passing from Eq. (B-14) to (B-15) is a consequence of the *Fredholm alternatives*. See, for example,

Courant, R., and D. Hilbert, *Methods of Mathematical Physics*, vol. I, Wiley-Interscience, New York, 1953.

Appendix C:
Basic Inequalities

C.1 Poincaré's Inequality

We shall prove the following theorem: *Let $u(z)$ be a bounded differentiable function in $0 \leq z \leq 1$ with $u(1) = 0$. Then*

$$\int_0^1 \left[\frac{du}{dz}\right]^2 dz \geq \frac{\pi^2}{4} \int_0^1 u^2 \, dz \tag{C-1}$$

The proof involves a trick. Let $h(z)$ be any differentiable function in $0 \leq z \leq 1$. Then

$$\int_0^1 \left[uh + \frac{du}{dz}\right]^2 dz \geq 0$$

This can be multiplied out and integrated by parts to give

$$\int_0^1 \left\{u^2 h^2 - u^2 \frac{dh}{dz} + \left[\frac{du}{dz}\right]^2\right\} dz + h(1)u^2(1) - h(0)u^2(0) \geq 0 \tag{C-2}$$

We now partially specify $h(z)$ so that

$$h(0) = 0 \qquad \text{(C-3)}$$

and we require that $h(1)u^2(1)$ remain finite. Then we can write Eq. (C-2) as

$$\int_0^1 \left[\frac{du}{dz}\right]^2 dz \geq \int_0^1 [h' - h]u^2 \, dz \qquad \text{(C-4)}$$

Finally, we take $h(z)$ as a solution to the equation

$$h' - h^2 = C^2, \qquad h(0) = 0 \qquad \text{(C-5)}$$

which is finite in $0 < z < 1$. The solution is

$$h(z) = C \tan Cz, \qquad C \leq \frac{\pi}{2} \qquad \text{(C-6)}$$

Thus,

$$\int_0^1 \left[\frac{du}{dz}\right]^2 dz \geq C^2 \int_0^1 u^2 \, dz$$

and the strongest inequality is obtained for $C = \pi/2$. It is readily established that

$$\operatorname*{limit}_{z \to 1} \tan \frac{\pi z}{2} u^2(z) = 0$$

as long as $u'(1)$ is finite. Thus, we obtain inequality (C-1), which is sometimes known as *Poincaré's inequality*.

By a nearly identical proof we prove the following theorem: *Let $u(z)$ be a differentiable function in $0 \leq z \leq 1$ with $u(0) = u(1) = 0$. Then*

$$\int_0^1 \left[\frac{du}{dz}\right]^2 dz \geq \pi^2 \int_0^1 u^2 \, dz \qquad \text{(C-7)}$$

C.2 Gavalas's Inequality

Let $u(z)$ be a bounded differentiable function in $0 \leq z \leq 1$ with $u(1) = 1$. Then

$$u^2(z) \leq [1 - z] \int_0^1 \left[\frac{du}{dz}\right]^2 dz \qquad \text{(C-8)}$$

The proof requires the Schwarz inequality,

$$\left[\int_a^b f(z)g(z)\, dz\right]^2 \le \int_a^b f^2(z)\, dz \int_a^b g^2(z)\, dz \qquad \text{(C-9)}$$

which is proved in most basic calculus texts. We write

$$u(z) = -\int_z^1 \frac{du}{dz}\, dz$$

$$u^2(z) = \left[\int_z^1 \frac{du}{dz}\, dz\right]^2 \le \int_z^1 \left[\frac{du}{dz}\right]^2 dz \int_z^1 1^2\, dz$$

$$= [1-z]\int_z^1 \left[\frac{du}{dz}\right]^2 dz$$

$$\le [1-z]\int_0^1 \left[\frac{du}{dz}\right]^2 dz$$

which is the required result. It is interesting to note that integration of inequality (C-8) gives

$$\int_0^1 \left[\frac{du}{dz}\right]^2 dz \ge 2\int_0^1 u^2\, dz$$

which is similar to, but weaker than, inequality (C-1).

C.3 Serrin's Inequality

Serrin's inequality is a multivariable generalization of Poincaré's inequality. We shall prove only a special case valid for cylindrical systems.

  Let $\mathbf{u}(r, z)$ be a bounded differentiable function which is periodic in z and vanishes at $r = R_1$ and $r = R_2$. $\mathscr{V}$ is a volume one period long in z between $R_1 \le r \le R_2$. Then

$$\int_{\mathscr{V}} \nabla\mathbf{u}:\nabla\mathbf{u}\, d\mathscr{V} \ge \left[\frac{\pi}{\ln(R_2/R_1)}\right]^2 \int_{\mathscr{V}} \frac{u^2}{r^2}\, d\mathscr{V} \qquad \text{(C-10)}$$

$u^2 = \mathbf{u}\cdot\mathbf{u}$. The proof parallels Sec. C-1. Consider

$$0 \le \|\mathbf{u}h + \nabla\mathbf{u}\|^2 \equiv [\mathbf{u}h + \nabla\mathbf{u}]:[\mathbf{u}h + \nabla\mathbf{u}]$$

$$= \nabla\mathbf{u}:\nabla\mathbf{u} + h^2 u^2 + \mathbf{h}\cdot\nabla u^2$$

$$= \nabla\mathbf{u}:\nabla\mathbf{u} + h^2 u^2 + \nabla\cdot\mathbf{h}u^2 - u^2\nabla\cdot\mathbf{h} \qquad \text{(C-11)}$$

We integrate Eq. (C-11) over the volume. The term $\int \nabla \cdot \mathbf{h}u^2 \, d\mathcal{V}$ becomes, by Green's theorem, the surface integral $\int \mathbf{n} \cdot \mathbf{h}u^2 \, d\mathcal{S}$. The surface integral vanishes at $r = R_1$ and $r = R_2$ by virtue of the boundary conditions. If we take $\mathbf{h}$ to be a function only of r, then the surface integral also vanishes along the z limits by virtue of the periodicity of $\mathbf{u}$, since the outward normal $\mathbf{n}$ points in opposite directions while $\mathbf{u}$ is unchanged. Thus, Eq. (C-11) becomes

$$\int_{\mathcal{V}} \nabla \mathbf{u} : \nabla \mathbf{u} \, d\mathcal{V} \geq \int_{\mathcal{V}} [\nabla \cdot \mathbf{h} - h^2] u^2 \, d\mathcal{V} \tag{C-12}$$

We now take $\mathbf{h}$ to be the bounded solution of

$$\nabla \cdot \mathbf{h} - h^2 = \frac{C^2}{r^2}$$

The solution is

$$\mathbf{h} = \frac{C}{r} \tan(C \ln r + D)\mathbf{i}_r \tag{C-13}$$

where $\mathbf{i}_r$ is the unit vector in the r direction. The strongest inequality is obtained when the argument of the tangent goes to $-\pi/2$ and $+\pi/2$ at the limits, in which case we obtain

$$C = \frac{\pi}{\ln(R_2/R_1)} \qquad D = \frac{\pi}{2} \frac{\ln(R_1 R_2)}{\ln(R_1/R_2)}$$

This value of C in inequality (C-12) gives the result, Eq. (C-10).

Appendix D:
Properties
of a Boundary
Value Problem

We shall develop here some general properties of solutions of linear second-order boundary value problems, starting with the special case

$$u'' + \psi(z)u = 0 \qquad \text{(D-1)}$$

$$u'(0) = u(1) = 0 \qquad \text{(D-2)}$$

This is an eigenvalue problem. $u(z) = 0$ is always a solution, but nonzero solutions will exist under certain conditions.

D.1 Lower Bound

We multiply Eq. (D-1) by $u(z)$ and integrate from zero to 1, integrating u'' by parts and using the boundary conditions. This gives

$$-\int_0^1 \left[\frac{du}{dz}\right]^2 dz + \int_0^1 \psi(z)u^2 \, dz = 0 \qquad \text{(D-3)}$$

The first integral is bounded by means of the Poincaré inequality (C-1), and we have

$$0 \leq -\frac{\pi^2}{4} \int_0^1 u^2 \, dz + \int_0^1 \psi(z)u^2 \, dz \tag{D-4}$$

Finally, we note that

$$\int_0^1 \psi(z)u^2 \, dz \leq \sup \psi(z) \int_0^1 u^2 \, dz$$

where $\sup \psi(z)$ denotes the least upper bound (supremum) of $\psi(z)$ in $0 \leq z \leq 1$. Thus, inequality (D-4) is further bounded by

$$0 \leq \left[-\frac{\pi^2}{4} + \sup \psi(z) \right] \int_0^1 u^2 \, dz \tag{D-5}$$

Since the integral of a square cannot be negative, $u(z)$ can be nonzero only if $\psi(z)$ satisfies the inequality

$$\sup \psi(z) \geq \frac{\pi^2}{4} \tag{D-6}$$

If the boundary conditions are changed to $u(0) = u(1) = 0$, then, by an identical proof, but using inequality (C-7), the corresponding result for a nontrivial solution is

$$u(0) = u(1) = 0: \quad \sup \psi(z) \geq \pi^2 \tag{D-7}$$

D.2 Sign Definiteness of $u(z)$

We show here that for

$$\frac{\pi^2}{4} \leq \sup \psi(z) \leq \pi^2 \tag{D-8}$$

a nontrivial solution $u(z)$ to Eqs. (D-1) and (D-2) will never equal zero in the interval $0 \leq z < 1$. The proof is by contradiction. Let ρ be a point in the interval $0 \leq z < 1$ where $u(\rho) = 0$. We define a new variable

$$\zeta = \frac{z - \rho}{1 - \rho}$$

Then Eq. (D-1) becomes

$$\frac{d^2u}{d\zeta^2} + [1 - \rho]^2 \psi u = 0 \qquad \text{(D-9)}$$

with boundary conditions

$$u = 0 \qquad \text{at } \zeta = 0, 1 \qquad \text{(D-10)}$$

According to inequality (D-7), this equation will have a nontrivial solution in the interval $0 \le \zeta \le 1$ only if ψ satisfies the inequality

$$[1 - \rho]^2 \sup \psi \ge \pi^2$$

Thus, for $\sup \psi < \pi^2$ the solution must be $u = 0$, $\rho \le z \le 1$. But if $u = 0$ for any finite interval $\rho \le z \le 1$, then $u = u' = 0$ at $z = 1$, in which case, according to the uniqueness criteria in Sec. 2.2, $u \equiv 0$ for all $0 \le z \le 1$. This contradicts the assumption of a nontrivial solution. Thus, we must have $\rho = 1$ as the only point in the interval where $u = 0$.

D.3 Eigenvalues and Eigenfunctions

We now turn to the eigenvalue problem

$$u'' + \psi(z)u = \lambda u \qquad \text{(D-11)}$$

$$u'(0) = u(1) = 0 \qquad \text{(D-12)}$$

The values λ_k for which nontrivial solutions exist are called *eigenvalues*, and the corresponding functions $u_k(z)$ are *eigenfunctions*. We shall prove two important properties of the eigenvalues and eigenfunctions:

1. Eigenfunctions $u_k(z)$ are orthogonal.
2. Eigenvalues λ_k are real.

Consider distinct eigenvalues λ_i, λ_j and the corresponding $u_i(z)$, $u_j(z)$. Then,

$$\int_0^1 u_i(z)u_j''(z)\, dz = u_i(1)u_j'(1) - u_i(0)u_j'(0) - \int_0^1 u_i'(z)u_j'(z)\, dz$$

$$= -u_i'(1)u_j(1) + u_i'(0)u_j(0) + \int_0^1 u_i''(z)u_j(z)\, dz$$

$$= \int_0^1 u_i''(z)u_j(z)\, dz \qquad \text{(D-13)}$$

All terms at the end points vanish because of the conditions (D-12). Substituting for u_i'' and u_j'' from Eq. (D-11) then gives

$$\int_0^1 u_i(z)u_j(z)[\lambda_j - \psi(z)]\, dz = \int_0^1 u_i(z)u_j(z)[\lambda_i - \psi(z)]\, dz$$

$$[\lambda_i - \lambda_j]\int_0^1 u_i(z)u_j(z)\, dz = 0$$

Since λ_i and λ_j are distinct, we get the orthogonality

$$\int_0^1 u_i(z)u_j(z)\, dz = 0, \qquad i \neq j \tag{D-14}$$

To prove that the eigenvalues are real we assume that both $u(z)$ and λ are complex,

$$u(z) = u_R(z) + iu_I(z) \qquad \lambda = \lambda_R + i\lambda_I$$

The real and imaginary parts of Eq. (D-11) can then be written as

$$u_R'' + \psi u_R = \lambda_R u_R - \lambda_I u_I \tag{D-15a}$$
$$u_I'' + \psi u_I = \lambda_I u_R + \lambda_R u_I \tag{D-15b}$$

It is readily established that

$$\int_0^1 u_R'' u_I\, dz = \int_0^1 u_R u_I''\, dz \tag{D-16}$$

Substituting Eqs. (D-15) then gives

$$\int_0^1 u_I[\lambda_R u_R - \lambda_I u_I - \psi u_R]\, dz = \int_0^1 u_R[\lambda_I u_R + \lambda_R u_I - \psi u_I]\, dz$$

$$\lambda_I \int_0^1 [u_I^2 + u_R^2]\, dz = 0 \tag{D-17}$$

Thus, $\lambda_I = 0$ and the eigenvalue is real. In that case the eigenvalues can be ordered, $\lambda_1 > \lambda_2 > \lambda_3 > \cdots$. It follows from inequality (D-6) that

$$\lambda_1 \leq \sup \psi(z) - \frac{\pi^2}{4}$$

Equations (D-11) and (D-12) are a special case of the *Sturm-Liouville equation*,

$$\frac{d}{dz}[r(z)u'] + \psi(z)u = \lambda p(z)u \tag{D-18}$$

$$a_1 u(0) + a_2 u'(0) = 0 \tag{D-19a}$$

$$b_1 u(1) + b_2 u'(1) = 0 \tag{D-19b}$$

where r, ψ, and p are real continuous functions in $0 \le z \le 1$, and $r(z) > 0$. By an identical proof it can be shown that the eigenvalues are real if $p(z)$ does not change sign in $0 \le z \le 1$. The eigenfunctions are orthogonal with weight p,

$$\int_0^1 p(z)u_i(z)u_j(z)\,dz = 0, \qquad i \ne j \tag{D-20}$$

Furthermore, the set of functions is complete, and any piecewise continuous function $f(z)$ in the interval $0 < z < 1$ can be expanded in a generalized Fourier series,

$$f(z) = \sum_{k=1}^{\infty} C_k u_k(z) \tag{D-21}$$

$$C_k = \frac{\displaystyle\int_0^1 p(z)f(z)u_k(z)\,dz}{\displaystyle\int_0^1 p(z)u_k^2(z)\,dz} \tag{D-22}$$

The series converges to $f(z)$ in the mean square sense.

BIBLIOGRAPHICAL NOTES

The material is covered in standard texts, such as

Courant, R., and D. Hilbert, *Methods of Mathematical Physics*, vol. I, Wiley-Interscience, New York, 1962.

Morse, P. M., and H. Feshbach, *Methods of Theoretical Physics*, McGraw-Hill, New York, 1953.

Weinberger, H. F., *A First Course in Partial Differential Equations*, Ginn/Blaisdell, New York, 1965.

Appendix E:
Adjoint Operator

For each differential operator there is a corresponding *adjoint* operator. Let

$$\mathscr{L}[\phi(z)] = 0 \tag{E-1}$$

be a linear homogeneous differential equation with homogeneous boundary conditions at $z = 0, 1$, and let $\phi^A(z)$ be any function. The adjoint operator $\mathscr{L}^A$ is defined as the operator with associated boundary conditions such that

$$\int_0^1 \phi^A(z)\mathscr{L}[\phi(z)]\, dz = \int_0^1 \phi(z)\mathscr{L}^A[\phi^A(z)]\, dz \tag{E-2}$$

It can be shown that the eigenvalues of the basic and adjoint systems are the same, though the eigenfunctions will be different in general.

As an example, consider the equation

$$\mathscr{L}[\phi(z)] = \phi^{IV} - \lambda\phi = 0 \tag{E-3}$$

$$\phi'(0) = \phi'''(0) = \phi(1) = \phi''(1) = 0 \tag{E-4}$$

We then form the integral

$$\int_0^1 \phi^A(z)[\phi^{IV}(z) - \lambda\phi(z)]\, dz$$

233

and, after four integrations by parts, using the boundary conditions (E-4), we obtain

$$\int_0^1 \phi^A(z)[\phi^{IV}(z) - \lambda\phi(z)]\, dz = \phi^A(1)\phi'''(1) + \phi^{A'}(0)\phi''(0)$$

$$+ \phi^{A''}(1)\phi'(1) + \phi^{A'''}(0)\phi(0)$$

$$+ \int_0^1 \phi(z)[\phi^{A^{IV}}(z) - \lambda\phi^A(z)]\, dz \quad \text{(E-5)}$$

We then satisfy Eq. (E-2) by taking the adjoint equation as

$$\mathscr{L}^A[\phi^A(z)] = \phi^{A^{IV}} - \lambda\phi^A = 0 \tag{E-6}$$

with boundary conditions

$$\phi^{A'}(0) = \phi^{A'''}(0) = \phi^A(1) = \phi^{A''}(1) = 0 \tag{E-7}$$

Note that the system (E-3) and (E-4) is *self-adjoint*, in that the operators $\mathscr{L}$ and $\mathscr{L}^A$ are identical with identical boundary conditions. An important property of self-adjoint systems is that the eigenvalues are always real. The proof in Appendix D.3 that the eigenvalues of the system (D-11), (D-12) are real exploits the self-adjointness of the system and can be generalized directly.

   Self-adjoint systems are not commonly encountered. Consider, for example, the Orr-Sommerfeld equation,

$$\mathscr{L}[\phi(z)] = \phi^{IV} - 2k^2\phi'' + k^4\phi - ik\,\text{Re}\,\{[1 - z^2 - c][\phi'' - k^2\phi] + 2\phi\} = 0 \tag{E-8}$$

$$\phi'(0) = \phi'''(0) = \phi(1) = \phi'(1) = 0 \tag{E-9}$$

By repeating the process carried out above the adjoint system is shown to be

$$\mathscr{L}^A[\phi^A(z)] = \phi^{A^{IV}} - 2k^2\phi^{A''} + k^4\phi^A - ik\,\text{Re}\,\{[1 - z^2 - c][\phi^{A''} - k^2\phi^A]$$

$$- 4z\phi^{A'}\} = 0 \tag{E-10}$$

$$\phi^{A'}(0) = \phi^{A'''}(0) = \phi^A(1) = \phi^{A'}(1) = 0 \tag{E-11}$$

   For systems of equations the approach is identical. The system is expressed as a matrix operator equation,

$$\mathscr{L}\cdot[\boldsymbol{\phi}(z)] = 0 \tag{E-12}$$

and the adjoint operator is defined from the equation

$$\int_0^1 \phi^A(z) \cdot \mathscr{L} \cdot [\phi(z)] \, dz = \int_0^1 \phi(z) \cdot \mathscr{L}^A \cdot [\phi^A(z)] \, dz \qquad \text{(E-13)}$$

BIBLIOGRAPHICAL NOTES

See, for example,

Greenberg, M. D., *Applications of Green's Functions in Science and Engineering*, Prentice-Hall, Englewood Cliffs, N.J., 1971.

Morse, P. M., and H. Feshbach, *Methods of Theoretical Physics*, McGraw-Hill, New York, 1953.

Author Index

Subject Index